FEATURES

SUMMER 2021·NUMBER 28

Plough

DEPARTMENTS

WEB EXCLUSIVES

plough.com/web28

Plough
PLOUGH.COM

EDITOR: Peter Mommsen
SENIOR EDITORS: Maureen Swinger, Sam Hine, Susannah Black
EDITOR-AT-LARGE: Caitrin Keiper
MANAGING EDITORS: Maria Hine, Dori Moody
POETRY EDITOR: A. M. Juster
DESIGNERS: Rosalind Stevenson, Miriam Burleson
CREATIVE DIRECTOR: Clare Stober
COPY EDITORS: Wilma Mommsen, Priscilla Jensen
FACT CHECKER: Suzanne Quinta
MARKETING DIRECTOR: Trevor Wiser
INTERNATIONAL EDITIONS: Ian Barth (UK), Kim Comer (German), Chungyon Won (Korean), Allen Page (French)
CONTRIBUTING EDITORS: Joy Clarkson, Leah Libresco Sargeant, Brandon McGinley, Jake Meador
FOUNDING EDITOR: Eberhard Arnold (1883–1935)

Plough Quarterly No. 28: Creatures: The Nature Issue
Published by Plough Publishing House, ISBN 978-1-63608-039-0
Copyright © 2021 by Plough Publishing House. All rights reserved.

EDITORIAL OFFICE
151 Bowne Drive
Walden, NY 12586
T: 845.572.3455
info@plough.com

SUBSCRIBER SERVICES
PO Box 8542
Big Sandy, TX 75755
T: 800.521.8011
subscriptions@plough.com

United Kingdom
Brightling Road
Robertsbridge
TN32 5DR
T: +44(0)1580.883.344

Australia
4188 Gwydir Highway
Elsmore, NSW
2360 Australia
T: +61(0)2.6723.2213

Plough Quarterly (ISSN 2372-2584) is published quarterly by Plough Publishing House, PO Box 398, Walden, NY 12586.
Individual subscription $32 / £24 / €28 per year.
Subscribers outside the United Kingdom and European Union pay in US dollars.
Periodicals postage paid at Walden, NY 12586 and at additional mailing offices.
POSTMASTER: Send address changes to
Plough Quarterly, PO Box 8542, Big Sandy, TX 75755.

Front cover: Aberdeen Bestiary, *The Panther*. Used by permission from Aberdeen University.
Inside front cover: Iris Scott, *Lunar Eclipse*. Used by permission. *irisscottfineart.com*
Back cover: Russell Powell, *Child of Light*. Used by permission. *pangaeanstudios.gallery*

ABOUT THE COVER: We took our cover imagery from the *Aberdeen Bestiary*, a twelfth-century English illuminated manuscript. Its detailed written and visual descriptions celebrate the abundance and variety of God's creations, from ant to elephant.

FORUM ≈

LETTERS FROM READERS

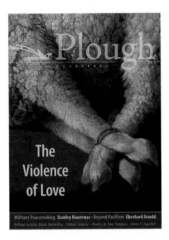

This Forum features selected responses to *Plough*'s Spring 2021 issue, "The Violence of Love"; for a fuller conversation, please see the digital version at *Plough.com/Forum27*.

The Forum is a place for commissioned responses by other writers to the questions raised by our authors, and for letters from you, our readers. Send contributions to *letters@plough.com*, with your name and town or city. Contributions may be edited for length and clarity, and may be published in any medium.

ALTERNATIVE TO WAR

On Scott Button's "A Life That Answers War": Reading this article as an Israeli brings to my mind two main questions: What is the price one is willing to pay for one's convictions? How is national interest defined?

Conscientious objection is not recognized in Israel. My father, who fought as a soldier in three wars, eventually refused to serve in a detention facility where humiliation of Palestinian detainees was common. Two of my brothers also refused to serve; all three spent time in prison. But there may be an even higher price for soldiers sent to control Palestinian civil populations or protect Israeli settlers in the occupied territories. A young Israeli told me the experience that most torments him from his army service is going into a Palestinian home in the middle of the night and having to watch over the wife and several children with his gun pointed at them. If more choose to see the humanity of Palestinians and refuse to serve the military occupation that oppresses them, our common interests will be much better defended.

Zohar Regev, Bethlehem, Palestine

Button's story of the Bruderhof and its cultivation of a different vision of conscientious objection is important because it offers an essential critique of how we resist war. Legal protections for conscientious objection moved during the twentieth century from those who were a part of a pacifist *tradition* to include *individuals* who object to war.

The Bruderhof vision preserves the notion that we are shaped as moral beings by our communities, and that our communities sustain us. Button's article shows that the American story of conscientious objection is the American story of war: moral actions governed by fragments of individual intuition, gut feelings, and appeals to political necessity. It is an American story in desperate need of an alternative.

To truly offer an alternative to war, we must begin not simply by emphasizing moral ideals like the Sermon on the Mount, but by being part of communities which help us to display, embody, and argue together about what those moral ideals might look like on the ground. In this, the alternative service of the objector is the extension of an alternative which has life even when the war ceases.

Myles Werntz, Abilene, Texas

NO PASSIVE PEACE

On Eberhard Arnold's "Beyond Pacifism": Arnold's commitment to the Sermon on the Mount pervades his whole theological outlook and generates a rich vision, addressing us from a time when Christian intellectuals, especially in Europe, were familiar with the work of Karl Marx and the plethora of socialist movements gaining traction across the globe. Arnold's theological criticisms of capitalism are as relevant as ever.

With the unyielding character of Arnold's nonviolent theology, however, there is a danger that we Christians – especially those who don't directly experience violence themselves – become detached from the realities of the oppressed. Writing in the 1920s and '30s, Arnold declares, "We must speak up in protest against every instance of bloodshed and violence, no matter what its origin." But

what of counterexamples such as the later Jewish uprisings in the ghettoes and concentration camps of Europe, in which the oppressed asserted their humanity against those who would have it annihilated?

Cameron Coombe, New Zealand

Here in Ethiopia, violence is the order of the day, and it is very common to hear of the killing of innocent people because of their ethnicity or religion, or for no clear reason at all. Some who call themselves freedom fighters have taken the law into their own hands, while the government seeks an end to violence, but not true peace. Pray for us! Suggest a solution!

Nonviolence or the absence of war does not represent peace. Peace is only perfect when it is eternal. Such peace is achieved only through love of God. In Jesus, we have reconciliation with others. He empowers us by his Spirit to be peacemakers with our neighbors, friends, and foes.

Birhanu Fanthun, Hawssa, Ethiopia

It is startling to find that Eberhard Arnold and Dorothy Day used the same words to describe our capitalist system: filthy, rotten. Both pointed to community as the solution: Eberhard Arnold held to complete community of goods, and the Bruderhof movement he founded has kept that rule. Dorothy Day and the Catholic Worker never aspired to that degree of community of goods. In our more modest moments we refer to "simplicity."

It is well to take a brief look at the social teaching of the Catholic Church, for Dorothy Day and other Catholic Workers have never conceived of themselves as anything other than faithful, if sometimes angry, sons and daughters of the Church. The basis of this teaching is Pope Leo XIII's 1892 encyclical *Rerum novarum*, which broke with powerful pressures within the Church by asserting workers' right to organize for collective bargaining and to strike. In 1931, Pope Pius XI issued *Quadragesimo anno*, a more explicit indictment of capitalism. While the Church recognizes the right to private property as a guarantor of liberty, that right is not absolute, and must be subordinated to the universal common good.

Tom Cornell, Marlboro, New York

In addressing the passivism of Tolstoy's Jesus, Arnold raises an important basic point: it is possible to misunderstand Jesus by taking his commands out of their larger context and not interpreting them with the whole of what God wills. From a magisterial Protestant point of view, this is unfortunately a problem that Arnold himself falls into.

Jesus' life was that of a private citizen with a specific and unique calling from God. While he is to be imitated in his character and the ways he commanded us to imitate him, he did not require us to have the same kind of life. That he did not marry does not mean his followers could not, and that he as a private citizen did not use violence does not mean Christians who are public officers may not.

Jesus' commands, which Arnold summarizes variously as love, justice, and peace, must be seen from a more comprehensive vantage point. Love, as desire for good, is the general principle of Christian living, but its application in context is more complicated than Arnold allows. Paul, who commended love as did his Lord, threatens the Corinthians that he will come "with a rod" if they do not repent. Anger is produced by injustice, and is sometimes the right thing for Christians to feel, but not always compatible with the gentleness and kindness that are the more "natural" fruits of love.

In some cases, arguably, Christians can engage in justified deception; in other cases, they may make a vow or take an oath for the sake of the peace and security of others.

Arnold focuses on God's revealed will for Christians *qua* Christians, and God's invisible rule in their souls through grace; he misses that God also rules through magistrates, sometimes through ecclesiastical judicial acts on the part of the visible church, and sometimes even through Christians with offices like that of magistrate, which come with exceptional divinely given obligations and permissions. Christians in such cases will still follow the basic principle of love for God, but while love is incompatible with malice towards others, it is not incompatible with causing them pain when necessary, though that necessary act will be attended with sorrow.

It would probably be a distortion to only spotlight disagreements with Arnold. And so magisterial Protestants must also say what needs to be said in his favor: he is to be commended in his zeal for justice, peace, love, and human flourishing. It seems to this writer beyond doubt that the individual lives of some Christians in the pacifist tradition (even if they might disclaim that label like Arnold did) are some of the most exemplary of Jesus' living Spirit on earth. If the world needs anything, it needs more of Jesus.

Andrew Fulford, Montreal, Quebec

DEFINING VIOLENCE

On Patrick Tomassi's "Behind the Black Umbrellas": I want to commend Tomassi on an excellent, even-handed article. I also want to state my position. I have spent years as a survivalist, a liberal, a Democrat, an Evangelical, and a poor person. I have known peace activists, Nazis, and Klansmen. Today I am an upper-middle-class older White guy who goes to a Roman Catholic church and identifies politically as a Christian Anarchist. I work as a street medic and care provider at BLM and antifa-led rallies and I have faced Proud Boys and their ilk in Washington, DC, prior to January 6. I am (striving to be) nonviolent.

I define violence as the intentional infliction of damage or prolonged pain on another person. Threat, coercion, and aggressive speech are all serious matters, but do not threaten bodily integrity. Property damage is often counterproductive, ineffective, or criminal, but not a threat to the person. Both sides use elastic definitions of words like *violence* and *self-defense* to justify themselves and demonize the other. That is why one of the key things we as Christians should try to do is define our words clearly and speak carefully.

While I have been to many antifa/BLM rallies, the only ones where I have seen violence occur were when Proud Boys or other counter-protesters showed up. I have seen many protesters carrying shields and umbrellas, but almost never anything I would term a weapon. Every time I have encountered Rightist protesters I have feared for my own safety. These are not equivalent groups.

At the same time, I must say that I am frequently disappointed with the refusal of self-analysis among BLM/antifa actors. I do not question the emotions that drive them, but I often wish that it did not evolve into a rejection of reason and a "race to the edge" to see who can express the most shocking "radical" opinion. Violence is a human problem, and will only be ended by changing human hearts. Antifa and BLM are at least actively committed to building loving, supportive, abundant (albeit secular) communities. I am not aware of any such efforts on the Right.

Brian Dolge, Catonsville, Maryland

Thank you for coordinating a diverse view in your issue from within the framework of Christian nonviolence. The violence-versus-nonviolence axis for understanding current events was acknowledged more frequently when the fault lines of cultural and political identity were more heavily aligned with foreign policy and antiwar activism. As our conflicts move onto more ambiguous battlefields – the "verbal violence" of memetic warfare in virtual space, the attempts to "cancel" or "doxx" public figures over controversies, the provocations of late-night protesters in improvised riot gear – it becomes harder to understand their boundaries. The focus on intent and outcome seems to swallow up all procedural definitions,

Continued on page 106

Vulnerable Mission in Action

Learning indigenous languages and prioritizing local funding, these missionaries follow a humble path of service, and help others do the same.

Jim Harries

I went to Zambia in 1988 in response to a call from God to serve people in the majority world. I was amazed at the level of misunderstanding I found between Western people living and working in Zambia and Indigenous African people there. All too often, confusion and error arose as a result of the enormous financial disparities between the two sides, which aggravated communication issues, even when all parties spoke English. In order to help resolve such issues, in 2007 I co-founded the Alliance for Vulnerable Mission, which advocates relating to people using their own languages and resources. I work closely with many Indigenous churches, teaching the Bible using the Luo and Swahili languages. I also look after orphaned children in my home in Kenya, where I have lived since 1993. I have chaired numerous conferences in the United Kingdom, Germany, and the United States, encouraging others to relate to African people in a vulnerable way, on their own territory.

The Alliance for Vulnerable Mission promotes mission outreach by those who are vulnerable, to people who are vulnerable. It believes that missionaries to the majority world should live and work vulnerably, humbly, from a position where they can listen. Listening and understanding requires use of the local language and dialects, as locals

Attendees take a break between sessions at a church conference in Kenya.

Dr. Jim Harries is the chairman and co-founder of the Alliance for Vulnerable Mission. He has lived in East Africa since 1988, teaching the Bible in Indigenous languages, caring for orphans, and working in hospital chaplaincy.

use them. It also means not contributing outside finances, which puts others in the position of having to talk for money. Instead, vulnerable missionaries work with local resources arising from family investment and home businesses, to contribute to the flourishing of locally rooted communities.

It is surprisingly difficult for Western missionaries to be vulnerable in Africa today. Supporters often press for quantifiable results: "How many people have you baptized?" or "How many wells have you dug?" Quantifiable results are much easier to achieve if you use outside money.

The use of an Indigenous language can make Bible interpretation complicated: Perhaps the word for God in one language means "source of prosperity," thus implying what the West thinks of as the "prosperity gospel." Perhaps the word for healing literally means "cooling," based on an understanding of disease as heat arising from tension in relationships or from witchcraft. The word for doctor might imply manipulation of mystical powers.

Missionaries have often tried to avoid such issues by using only English. They have used donations to cancel opposition. These actions have made majority-world people dependent on unfamiliar, outside ways of bringing

---

For more on Jim's story, read his novels African Heartbeat *and* To Africa in Love, *both published by Faithbuilders Christian Books. Find out more at* vulnerablemission.org.

both "prosperity" and "cooling." Vulnerable missionaries seek to walk the Jesus way, to go where they can share the gospel from within the complexities of Indigenous ways of life. If local means seem slow or weak by comparison with foreign money and resources, we say with a wise missionary of long ago, "When I am weak, then I am strong" (2 Cor. 12:10).

Regenerative Agriculture in Austria's Weinland

The Bruderhof in Retz, Austria, runs a business that fits into the local economy, sells weekly vegetable boxes, and fosters care of the creation – soil, plants, worms, even microbes. **Plough's Andrew Zimmerman finds himself digging into the soil-friendly principles of intensive market gardening – and broadforking beds of brassicas.**

Andrew Zimmerman

I was never much of a gardener. Sure, I put a few cherry tomatoes in a small plot each summer, along with some basil and parsley. But I had never heard of the five principles of regenerative agriculture, and I was certainly not aware that broccoli, kohlrabi, and kale are essentially the same plant, Brassica oleracea, that has been selectively bred over centuries.

After moving to Austria two years ago to help found a new Bruderhof community, that changed. To earn our living, we wanted to find a business that fit in with the local economy. Agriculture was one thing that made sense; the local

landscape was dominated by vineyards, fields of sunflowers, and pumpkin patches, and the property we moved into had a seven-acre field attached.

But we didn't want to go into industrial farming. One of our community members had studied organic agriculture in Germany, and was convinced of the dangers of pesticides and herbicides. Plus, in a dry climate – we receive only around fifteen inches of rainfall annually – water was going to be crucial.

So we turned to regenerative agriculture, and started an intensive market garden on one of the acres. The core principles of this type of farming call for minimal soil disturbance, maximum crop diversity, keeping the soil covered with vegetation, integrating livestock, and leaving old rootstocks in the ground.

Planting many different kinds of crops isn't too hard, and actually helps our business model, since we sell weekly vegetable box subscriptions and participate in local farmers' markets. Integrating livestock? We spent the winter building a mobile stall for our new free-range laying hens.

Some of the other principles are more work: keeping the soil covered reduces wind erosion and water evaporation, but to achieve that, you can't sow seed directly. So we start plants in our greenhouse and then have weekly transplanting projects of several thousand "plugs." While the work is almost literally backbreaking (at least for me, a tall guy in his mid-forties) it involves our entire twenty-odd community members, and connects me with the earth in a way

Image courtesy of Andrew Zimmerman

Selling garden produce in the nearby town of Retz

that my other work (posting behind-the-scenes videos on Facebook or dealing with customers who want to skip a box when they go on vacation) doesn't.

Minimizing soil disturbance means that after harvesting a bed, we don't plow it, which would expose and reinvigorate dormant weed seeds. Instead, we loosen the soil with a broadfork and leave the roots there to decompose. This keeps organic matter in the ground where it belongs, and drastically reduces weed problems.

We've had a great response. The box subscription program is taking off – the pandemic may have helped us, as many customers want home delivery – and we've doubled the number of beds under cultivation for Year Two.

But to me it's bigger than just agricultural principles and a successful startup. It is care of the creation – soil, plants, worms, even microbes – that God has given us, as in Genesis 1:29. Not to sound grandiose, but what we're doing is saving the planet – at least our little corner of it.

I'm still not much of a gardener, but I get my hands dirty a couple times a week. Bruderhof founder Eberhard Arnold wrote that "a person's intellectual life needs to be stimulated and deepened, if he is to become human. But only when he knows the pleasures of physical labor will he experience the joy of being – joy in God and joy in his creation." I couldn't agree more. Sure, I'm no match for my teenage sons, who can broadfork a twenty-yard bed in about three minutes, but I know my brassicas now.

Poet in This Issue

Alfred Nicol collaborated with Rhina Espaillat and illustrator Kate Sullivan to create the chapbook *Brief Accident of Light: Poems of Newburyport* (Kelsay Books, 2019). Nicol's most recent full-length collection of poetry, *Animal Psalms*, was published in 2016 by Able Muse Press. He has published two other collections, *Elegy for Everyone* (2010), and *Winter Light*, which received the 2004 Richard Wilbur Award. His poems have appeared in *Poetry*, the *New England Review*, *Dark Horse*, *Commonweal*, *The Formalist*, *The Hopkins Review*, *Best American Poetry 2018*, and many other literary journals and anthologies. Nicol lives in West Newbury, Massachusetts, with his wife, Gina DiGiovanni.

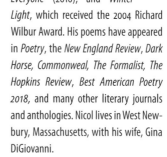

See his poems "The Path" on page 43, and "The Berkshires" on page 61.

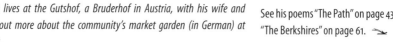

Andrew Zimmerman lives at the Gutshof, a Bruderhof in Austria, with his wife and three children. Find out more about the community's market garden (in German) at gutesvomgutshof.at.

The Work of Thy Fingers

Singing God's Glory with Keith Green

RANDALL GAUGER

AROUND 1978, as new Christians, my wife, Linda, and I discovered some music that played an important role in our seeking to follow Jesus. Keith Green's music was different from the praise and worship music we were familiar with. His songs about repentance, the costs of discipleship, and the struggle to give everything to Jesus were both heartfelt and hard-hitting. We were drawn to both the music and the man.

Born in 1953, Keith was about my age. He had been a child actor, a preteen singer-songwriter, and a late-sixties hippie. In 1975 he converted to Christianity, along with his wife, Melody. Once that happened he never turned back. In 1979 the Greens moved to Texas to start a community called Last Days Ministries. Keith continued to write songs while also ministering to people in need, including the homeless and pregnant women.

Though Keith died in a plane crash in 1982 at age twenty-eight, his music continues to speak to listeners today. Many of his songs are scripture put to music. I especially love his 1982 setting of Psalm 8:

Oh Lord, our Lord,
How majestic is thy name in all the earth.

When I first listened to Green's version of the psalm, it struck me in a new way with the truth of God, his creation, and our place in it. The glory of nature echoes the glory of God:

For when I consider the heavens
  the work of thy fingers.
The moon and the sun which thou
  hast appointed.
For what is man, that thou
  dost take thought of him,
And the son of man, that thou
  dost care for him?

Psalm 19 talks about the heavens too: "The heavens are telling the glory of God; and the firmament proclaims his handiwork." Nature itself reveals truth and knowledge to us, and every creature tells us something about its creator. As G. K. Chesterton says in his 1925 book *The Everlasting Man*:

What we know, in a sense in which we know nothing else, is that the trees and the grass did grow and that a number of other extraordinary things do in fact happen; that queer creatures support themselves in the empty air by beating it with fans of various fantastic shapes; that other queer creatures steer themselves about alive under a load of mighty waters; that other queer creatures walk about on four legs, and that the queerest creature of all walks about on two.

By observing nature, both in the heavens and on the earth, we see God's invisible qualities, his eternal power and majesty. I have often stood in awe gazing at the star-flecked night

---

Randall Gauger, a pastor in the Bruderhof communities, lives with his wife, Linda, in upstate New York.

Photograph credits: NASA, ESA, H. Teplitz and M. Rafelski (IPAC/Caltech), A. Koekemoer (STScI), R. Windhorst (Arizona State University), and Z. Levay (STScI)

sky, and the more I learn about the universe, the more I discern the hand of a divine creator.

> *Hallelujah Lord our Lord,*
> *Oh how I love you.*
> *You've made man a little lower*
>   *than all of the angels,*
> *And crowned him with glory to rule*
>   *over your creation.*

Under that night sky, I have echoed the words of King David: "What are human beings that you are mindful of them, mortals that you care for them?" Contemplating such a great expanse, I find God's attention and dedication to us tiny humans astounding.

NOT ONLY DID GOD create us and crown us, he made us his image-bearers. It is this that gives human beings their unique place in nature. After he created the earth, God did not become a passive viewer. He maintained connection with his creation and in particular with humankind, loving and cherishing each human being. He gave humanity the task to nurture his creation, to create new life by being fruitful but also to care for and protect his masterpiece so it would remain a place where all creatures could thrive.

Despite God's commission, humans do not do such a great job of creating that place of joy and peace on earth. So God does something amazing and wonderful to call us back to his original intent: he sends Jesus. God steps out of himself and becomes one of us. He does this out of his infinite love for us, demonstrating that there is no length to which he will not go out of love for each of us. Throughout his ministry, Jesus calls his listeners back to God's will for creation, urging his followers to heal rather than harm, nurture rather than ravage, and make peace rather than war.

Jesus promises that he will come again and usher in the kingdom of God, the new and perfected creation. We will have new bodies that will never get sick and die. As it says in Isaiah 65: "For I am about to create new heavens and a new earth; the former things shall not be remembered or come to mind." Peter picks up this refrain in his second letter: "That day will bring about the destruction of the heavens by fire, and the elements will melt in the heat. But in keeping with his promise we are looking forward to a new heaven and a new earth, where righteousness dwells." And these promises echo once again in the prophetic words of Revelation:

> Then I saw a new heaven and a new earth; for the first heaven and the first earth had passed away, and the sea was no more. And I saw the holy city, the new Jerusalem, coming down out of heaven from God, prepared as a bride adorned for her husband. And I heard a loud voice from the throne saying, "See, the home of God is among mortals. He will dwell with them; they will be his peoples, and God himself will be with them; he will wipe every tear from their eyes. Death will be no more; mourning and crying and pain will be no more, for the first things have passed away." And the one who was seated on the throne said, "See, I am making all things new."

I hope you take the time to listen to Keith Green singing Psalm 8, expressing the beauty of God's creation and the gift of being image-bearers. Our God is involved in creation and in our lives. He has given us work to do on earth as we await the new heaven and new earth that Jesus will bring. God has placed everything into the hands of Jesus. No matter what circumstance we find ourselves in, this psalm lifts our hearts and eyes to God, his creation, and the best that is yet to come.

> *Hallelujah Lord our Lord,*
> *Oh how I love you.*

Hubble Ultra Deep Field, 2014

The Book of the Creatures

PETER MOMMSEN

Artwork from the *Aberdeen Bestiary,* a twelfth-century illuminated manuscript

DOGS EVOLVED "expressive eyebrows" to trigger feelings of affection in humans, according to a 2019 study reported in *Proceedings of the National Academy of Sciences.* The researchers found that in the thirty thousand years since dogs separated from wolves and began consorting with us, their faces have changed so that their eyes "appear larger and more infant-like" and are capable of mimicking human expressions. When they look at us, we feel the same tenderness as when we're face-to-face with a young child. Put more cynically, dogs have managed to hack into our most primal emotion.

Is it then instinctive manipulation when my Brittany hound gazes at me with his sad and eager eyes? No doubt, but that's not the whole story. By analyzing hormone levels, the same study showed that dogs feel a pleasurable rush when their masters show them affection. Their masters feel the same, thanks to the same chemical, oxytocin. Evidently, we have learned to communicate as fellow creatures who genuinely enjoy each other's company.

People in earlier centuries took the joy we feel in other living things at face value, as a pointer to a theological truth. "All things bright and beautiful, all creatures great and small, all things wise and wonderful, the Lord God made them all," wrote the Anglo-Irish poet Cecil Frances Alexander in her well-known 1848 hymn. With childlike sweetness, the hymn sums up a core belief shared by most religions. Flowers, birds, humans, stars: we are all creatures, the handwork of a Creator. "In the beginning God created heaven and earth," declares the first sentence of Genesis, which then sums up the creation story with the affirmation: "And it was very good."

This very-goodness will be familiar to anyone, religious or not, who loves nature. It's a spontaneous wonder at the beauty – the apparent *meaningfulness* – of life. It's the inkling one gets when hiking, let's say, through

Image from *Aberdeen Bestiary.* Used by permission from Aberdeen University.

a forest on a cool morning in early summer, hearing a red-winged blackbird sing in the sumac, watching largemouth bass dimpling the surface of a misty lake, startling a fawn up from the brush. All this strikes one powerfully as so obviously good that it seems to suggest a Goodness behind it all (page 62).

This intuition is an ancient one common to traditional cultures the world over, whether Aboriginal Australian, Shinto, or classical Greek. In Christianity, a long tradition going back through Galileo and Bonaventure to the third-century Desert Fathers holds that the natural world is a book that reveals the divine, just as the book of scripture does. Nature is legible. As the Egyptian hermit Saint Anthony, who spent decades living in the wilderness, expressed it: "My book is the created nature, one always at my disposal whenever I want to read God's words."

When we read the book of nature, what is it that we read? Nothing less than a description of who God is, insisted this tradition. In the words of Basil of Caesarea: "We were made in the image and likeness of our Creator, endowed with intellect and reason, so that our nature was complete and we could know God. In this way, continuously contemplating the beauty of creatures, through them as if they were letters and words, we could read God's wisdom and providence over all things."

In Basil's view, reading the book of nature is mostly a matter of contemplation: paying attention, not analyzing. Unlike a scientific researcher, a contemplative doesn't ask first what things are made of or how they function. Instead, she aims to commune with the goodness with which the Creator has suffused creation, a goodness that can heal troubled human hearts. Reading the book of nature is therapeutic. (This same insight still drives today's efforts to reconnect people with nature, from national park systems to Fresh Air programs for city dwellers, from Scandinavian forest schools to Japanese "forest bathing.")

It's hard to overestimate the value that early Christians placed on the right appreciation of nature. This isn't surprising, since the sayings of Jesus himself brim with delight in the "birds of the air" and the "lilies of the field." Jesus' love of the natural world, in turn, reflected that of Hebrew scriptures, especially the Psalms:

> The heavens are telling the glory of God;
>> and the firmament proclaims his handiwork.
> Day to day pours forth speech,
>> and night to night declares knowledge. . . .
> Their voice goes out through all the earth,
>> and their words to the end of the world.

The creation's "words" thus communicate to everyone, everywhere. Augustine of Hippo, one of the foremost teachers of scripture, judged nature's testimony to be superior to the Bible in at least one respect: it is accessible to everyone, even those who cannot read or write. "There is a great book: the very appearance of created things. Look above and below, note, read. God, whom you want to discover, did not make the letters with ink; he put in front of your eyes the very things that he made. Can you ask for a louder voice than that?"

When we forget how to read nature, we forget how to read ourselves.

BUT THE BOOK OF NATURE reveals many things that are less lovely than stars in the sky and singing birds, a side of creation famously described by Alfred Lord Tennyson as "red in tooth and claw." This account of nature found its most influential articulation in Charles Darwin's 1859 *Origin of the Species.*

To be clear, Darwin's theory in itself doesn't necessarily conflict with faith (page 79), as theologians from Karl Barth to Benedict XVI have pointed out. (The particular brand of biblical literalism that it does exclude was already rejected by Augustine and Origen.) In fact, evolution claims a majesty of its own, as Darwin's book concludes:

> There is grandeur in this view of life, with its several powers, having been originally breathed into a few forms or into one; and that, whilst this planet has gone cycling on according to the fixed law of gravity, from so simple a beginning endless forms most beautiful and most wonderful have been, and are being, evolved.

Yet it's understandable that to Cecil Frances Alexander's generation Darwin's revolution came as a shock. In his account, nature is not primarily bright and beautiful. Instead, it is a world of desperate competition for survival, of mass extinctions and genetic dead ends, of disfiguring diseases and cruel parasites. Is such a world a convincing argument for the Creator's goodness? Many have concluded that it is not. As the British comedy troupe Monty Python put it in their 1980 rewrite of Alexander's lines: "Each little snake that poisons, each little wasp that stings, He made their brutish venom, He made their horrid wings." Even if a deity had created such a world, skeptics charge, what kind of deity would he be?

It's a legitimate challenge – one with a special resonance as a pandemic continues to rage. Monty Python's satire may not mention contagious respiratory viruses. But the coronavirus microbe fits neatly into their song's catalog of "all things sick and cancerous, all evils great and small."

Put another way: How can evil exist in a good creation? Darwin was hardly the first to notice the existence of natural evil; it's already addressed in the Book of Job, likely the oldest book in the Hebrew Bible. This question has also had a long history of Christian reflection, going back at least to the apostle Paul. In his Letter to the Romans, he wrote that creation is "subjected to futility" and in "bondage to decay," "groaning in labor pains until now." It's a passage that seems remarkably apt as a description of the realities of natural selection – or indeed, of the Covid pandemic.

The answer to the riddle, for the early Christian authors, lies in the nature of reality itself. All creatures, they believed, are words in the book of nature; but that book's preeminent Word is the Logos, the Word made flesh. This Word is the grand theme of the opening chapter of John's Gospel – not coincidentally Basil of Caesarea's favorite text. Here, the Word who made all things enters into the futility of his groaning creation and through his own death and resurrection rescues it from its evil and suffering. In the pithy formulation of another church father, Maximus the Confessor, when we read the book of nature, what we are really reading is "the words of the Word."

THIS KIND OF READING doesn't come easy to us moderns. Why do we have trouble seeing what seemed clear to these early authors? One reason is straightforward: we're simply out of practice. This is partly a result of rapid urbanization – today for the first time in human history, the majority of people don't live out their lives surrounded by the natural world. Unfamiliarity leads to estrangement. Children, especially, are missing out on the encounters with nature that were formative for earlier generations. In 1920, 30 percent of Americans lived on a farm; today only 1 percent do. Over the past fifty years, the share of Americans who hunt has halved, so ever fewer kids learn the ways of wildlife by heading into the woods. And 80 percent of Americans can't see the Milky Way from where they live.

Instead of lived experience and childlike reception, we moderns are more likely to approach nature with the analyzing bent of modern science. This in itself is an excellent means of unlocking it: studying ornithology only increases the birdwatcher's sense of wonder; knowledge of river biology enhances the fisherman's love of the sport. More broadly, modern science's analytic tools command wonder in their own right – for example when geneticists discover Neanderthal genes in modern human DNA, or physicists identify strange behavior by muons that may give hints about the 95 percent of the universe that we know almost nothing about because it consists of dark matter and dark energy.

All the same, the empirical data supplied by the scientific approach are only one kind of knowledge, and this is what modernity tends to forget. As C. S. Lewis observed, there is a difference between knowing what a star is *made of,* and what a star *is.* Today we know more by several orders of magnitude than Basil or Maximus did about what nature is made of and how it works – even what gives a dog his puppy eyes. But that does not yet tell us what nature *is* or what nature is *for.*

This illiteracy applies to human nature as well. When we forget how to read the book of nature, we forget how to read ourselves. The givenness of our humanity no longer yields self-evident truths about what rights or purposes our Creator may have endowed us with, or indeed about who we really are. What then is there to guide and limit the bounds of technological manipulation of humankind? Transhumanism may remain a fantasy, but CRISPR gene editing of babies is not (page 80). Meanwhile, our behavior untethered from any understanding of the natural good has wreaked havoc on life forms across the planet – a careless destruction that will ultimately harm us too.

This is where the ecological movement can perhaps ride to the rescue. For the one great gain arising from our civilization's crime of global environmental destruction is the growth of ecological consciousness. We're slowly – far too slowly – realizing what it would mean to

lose the natural world from which we arose and to which we belong.

To be sure, environmental activists bolster their cause by citing statistics showing the costs of climate change in habitat loss, economic and social disruption, or perhaps the disappearance of specific species. But underlying such rhetorical tools is a more basic conviction: That nature in all its intricate diversity has an inherent dignity and beauty. That ecosystems and landscapes are goods in themselves that ought not to be destroyed, regardless of any effects on GDP. That blue whales, African elephants, and the Amazon rain forest have a value of their own that we are bound to respect.

Here then is an example of millions of people worldwide re-learning to read the book of nature, even if only partially and imperfectly. Perhaps as we learn to read again, we'll find that human nature too becomes more clearly legible. And perhaps too, as we read more deeply, we'll again learn to decipher the signs of the Goodness behind nature – the one who is both author and subject of the book of the creatures, and who, in Alexander's words, "gave us eyes to see them, and lips that we might tell how great is God Almighty who hath made all things well." ⇘

Return to Idaho

GRACY OLMSTEAD

STEP INTO A FAMILIAR GRAVEYARD. My husband is with me, and so is our oldest daughter, a curly-haired girl with dark lashes and thoughtful eyes. I am showing them the graves of my forebears, here where generations of my family rest in the soil of Emmett, Idaho.

I brush snow off gravestones, searching for names. There are my great-grandparents, known as Grandpa Dad and Grandma Mom,

---

Gracy Olmstead is a journalist whose writing has appeared in the American Conservative, *the* Week, *the* New York Times, *and the* Washington Post, *among others. Her most recent book is* Uprooted: Recovering the Legacy of the Places We've Left Behind *(Sentinel, 2021), from which this essay is adapted.*

When I went back to Idaho, I connected with more than just the land.

home in this quiet land, where purple shadows line the foothills and the Payette River is shadowed by cottonwood trees. It is good to be here, in a place where the past is still present and preservation is paramount. But much has changed here since I was a child.

Fields that once were filled with corn and sugar beets, mint and onions are now graded, leveled, and covered with single-family homes. The farm stand where we bought tomatoes and peaches for canning is gone – as are many of the small mom-and-pop businesses we frequented growing up. Everywhere I drive in my homeland, I see the past crumbling and fading away, increasingly paved over and forgotten. And even as I observe good changes, I mourn what's been lost and what we are losing.

I grew up surrounded by folks who committed themselves to this place for the long haul. They served and loved it, year after year. For many of them, including my great-grandparents, rootedness meant turning down bigger paychecks, adventure, excitement, and ease. But they were able to experience the wonder of each season, the pride of committed stewardship. They got to grow old next to the ones they loved – and got to watch a new generation of young folks, like me, grow up in the land they had tended.

Wallace Stegner once called the United States' two archetypal populations the "boomers," who come to extract value from a place and then leave, and the "stickers," those who settle down and invest. Now boom-and-bust cycles and the exodus of the young, including me, have worn down the threads of community and belonging. Many of the valley's rooted individuals are still alive – but they are growing old.

I learned my first stories about this town at the feet of my Grandpa Dad, who was an aged farmer by that point. He knew all his neighbors, and they knew him. He was out riding

Bob Bales, *Emmett Valley farmstead*, 2017

side by side in the earth. Nearby, I see the graves of Grandpa Dad's brothers, his sister, and his parents – along with dozens of other, more distant kin. My husband holds our daughter as he quietly reads inscriptions.

To visit these graves today, I had to travel twenty-four hundred miles: over the Blue Ridge Mountains and the Great Plains, past the Rockies and the Sawtooths. It is good to be

his tractor until his early nineties, attending church services and teaching Sunday school until the end of his days. Over his lifetime, this valley aged into post-industrialism and decay.

"Grandpa Dad," Walter Howard on his farm, ca. 1978

That corn planted roots deep inside me, connecting me to Grandpa Dad and to the land that he cared for.

I am sure that much of what he saw alarmed him, inspiring in him a sense of sorrow for the land's lost past, its fading local culture. I think that is why he told me stories over and over again, making sure I would not forget the ones who came before us, the countryfolk that history books would ignore. To Grandpa Dad, their stories mattered. And so they mattered to me as well.

There was a time when the Drug Enforcement Administration showed up at Grandpa Dad's door, demanding to know what he was growing in the middle of his cornfield. They had seen an aerial view of the field and knew that he had planted something in the middle of it that was shorter than the field corn surrounding it. They suspected this snowy-haired old man might be growing marijuana.

Grandpa Dad assured the concerned agents that he was doing nothing of the sort. He had started a tradition some years back of planting sweet corn – which is about two feet shorter than field corn – in the middle of one of his other cornfields. He didn't want to grow the sweet corn right up to the road's edge because passersby would occasionally stop their cars and help themselves to the crop. (Yes, locals in farm country can tell the difference between the full leaves and tall height of field corn and the shorter, spindlier appearance of sweet corn.)

The sweet corn that Grandpa Dad planted wasn't for him, you see – it was for church folks, neighbors, and family. So Grandpa Dad would plant it a little ways inside the tall rows of field corn, set back just enough from the road to hide it from prying eyes.

The field of hidden sweet corn was Grandpa Dad's first fruits: the crop he grew to give away. It was the corn I shucked with clumsy hands as a little girl and ate all through the cold winter months. It prompted storytelling and feasting, the gathering of the generations to bring in the harvest. That corn planted roots deep inside me, connecting me to Grandpa Dad and to the land that he cared for.

I can imagine Grandpa Dad planting that field in the spring, memorizing poetry and scripture verses as he worked on his tractor. I'm sure he looked forward to seeing the green tendrils of life erupt from the earth, to the truckloads of corn he would drive to my grandparents' house, to the laughter and music we would all enjoy together. My ancestors were always setting aside a portion of their proceeds to bless the people they loved. Their labor was never just for them: it was a poured-out thing.

Over the past several years, I've learned that the dead can hurt or heal, urge us forward or call us back. This means that the work of the boomers, those who deplete soil and community, can result in long-lasting brokenness. It can take generations to recover from their legacy, to restore what's been depleted. Many of us are still waiting and watching, hoping to see

our homelands restored. We are still observing and mourning what has been lost, squandered, or abused.

But stickers, in contrast, can sow blessings in the soil for decades to come. I owe much of the good fruit in my life to my ancestors' lives, labors, and love: to the chain of membership they handed down, the values they passed on, the richness they built in their community. My forebears connected me to much more than the land: They connected me to the dead, who came before me, and to the seasons that surrounded me. They connected me to rhythms of family, community, and virtue that sprang up in this land long before I was born. The portion they set aside resulted in an overflowing abundance of joy and grace.

In his work *Reflections on the Revolution in France*, Edmund Burke suggests that society is not just made up of the living but serves as an association between the dead, the living, and the unborn. To be indebted is to see oneself as inseparably intertwined with the duties and responsibilities of this membership. We are never entirely solitary or self-determining in this life. Everything we have and are is inescapably tied to those who came before us.

That doesn't mean that we cannot make our own marks on this world, that we cannot forge new paths for ourselves or for our families. But it acknowledges the fact that we are part of a community both dead and alive, and that this reality comes with responsibilities. We Americans delight in seeing ourselves as self-made, as mavericks. Tocqueville was right about us. We don't want to acknowledge what we might owe to the past or to place. But I think Wendell Berry is also right: we should take our membership seriously, considering those dead and alive who have made us who we are, and how we might further their work in the future. The past is never fully past – not for the soil, and not for us. ⤳

Bob Bales, *Emmett Valley,* 2017

ADAM NICOLSON

Into the Weald

In my corner of England, the labors of people
long dead live on in the landscape.

I F YOU KNEW the Weald of Sussex only from looking at its place on a map of England, you would never guess what it was. Squeezed between the outskirts of London to the north and the rim of coastal towns to the south, heavily crossed by trunk roads and railways, with some sizable towns within it, plus a giant international airport, what would it look like but a slab of post-industrial suburbia, a place whose deeper meanings were hidden under a busy, scarred, modern, utilitarian surface?

Even when you drive through it, met at every turn with the signboards that line the main roads, you would not get to feel what it was. No, there is only one way and that is on foot, away from traffic and its noise, allowing quietness and a sense of the past to come seeping up out of the ground beside you.

Choose one of the many hidden and sunken lanes that crisscross this damp and tree-thick country. Any large-scale map will suggest them but choose one that sinks down into an unvisited valley. Immediately the place closes over you. Often the clay walls of the lane are moss-lined, in spring wallpapered with primroses and wood anemones, pushing down through the bluebells and woodspurge, coming down to a little tree-roofed stream that is creased into the clays and sandstones of the country.

Much of that deep, damp, wooded lowland of the Weald is almost abandoned now, the coppiced hazels, hornbeams, and ashes neglected for decades so that each stem from their ancient stools is now a full-grown forest tree, which here and there has crashed out in winter gales to lie horizontal among its neighbors.

These woods – and this is the most wooded part of England – grow steadily darker as the summer thickens, and that beautiful concealed cool of midsummer shade is what Rudyard Kipling loved in the valley of the river Dudwell near Burwash where he lived:

Even on the shaded water the air was hot and heavy with drowsy scents, while outside, through breaks in the trees, the sunshine burned the pasture like fire. . . .

The trees closing overhead made long tunnels through which the sunshine worked in blobs and patches. Down in the tunnels were bars of sand and gravel, old roots and trunks covered with moss or painted red by the irony water; foxgloves growing lean and pale towards the light; clumps of fern and thirsty shy flowers who could not live away from moisture and shade.

Kipling came here after an adventurous life in India, America, and South Africa, looking for a place in which he could exchange vast imperial

Opposite:
Gill Bustamante, *Epiphany,* impasto paint on canvas, 2018

Linoprints by Carolyn Cox

Adam Nicolson is an author and journalist who has written for the Sunday Times, *the* Sunday Telegraph, National Geographic, *and* Granta, *where he is a contributing editor. He is the author of numerous books on landscape, literature, history, and the sea, most recently* The Making of Poetry: Coleridge, the Wordsworths, and Their Year of Marvels *(Farrar, Straus and Giroux, 2020). He is married to the writer and gardener Sarah Raven and lives at Perch Hill Farm in Sussex.*

Gill Bustamante,
Celebration,
oil on canvas,
2020

distances for something narrower, deeper, and older, a swapping of the giant horizon for the deep well of ancientness. He could not have chosen better.

I do not know of a landscape that is so full of the suggestions of the past. This may be because modernity has abandoned so much of it, allowing the woods to return to a wildness they have probably not had since the Dark Ages, leaving the old handmade structures to moulder – not only the lanes, but the woodbanks that once protected the young trees from the deer, the coppice stools themselves, the interfolding of field and wood, the remote abandoned ponds once dammed for ironworks, the pits where the farmers dug down for the limy marl to sweeten their acid soils.

I first came to live here thirty years ago, on the run from London and looking for a refuge, a place where my wife, Sarah, and I could make a life. We found a small, run-down dairy farm, out in the rather rough country on the borders of Brightling and Burwash. It was called Perch Hill, meaning stick, or pole, hill – the high clay soils of the fields were good for growing little but the coppiced trees. It was the best

decision of my life: the poverty of the farm had preserved most of its essential forms, its woods and fields. All we had to do was strip out the concrete and corrugated iron, put back some of the hedges, plant some trees, and find a rich and precious place gathering around us.

It is in large part a deserted country. Mechanization and the dominance of distant markets mean that it no longer needs to be occupied as it was when this landscape was made, in the great surge of population in the twelfth and thirteenth centuries, when the weather was better, there was no frost in May, and the people here were pushing out even into the marginal land that the Weald has always represented.

You would have to go to an equally poor, undeveloped, and still handmade country such as the valleys of eastern Transylvania to find an atmosphere alive today that in any way resembled the historic Weald. I had read about the wonderful Transylvanian meadows, not yet destroyed with herbicide or too much nitrogen fertilizer, and so one spring went with Sarah to look for the flowers. What I hadn't expected was to find a country that was inhabited in ways that suddenly struck all kinds of echoes with what must once have been life in the Sussex Weald. The forms are strangely the same – wooden farmsteads, hedged fields, the cultivation of the meadow, a place of cattle, the well-used forest – but the most surprising part of Transylvania for a Sussex man is just how busy it is in every corner: in early spring, when the snow has gone, the whole world around you there is filled with people plowing, hoeing, axing out the dead wood from the pollards, leveling molehills, cutting bean sticks, planting beans, raking old leaves, putting out dung. Women walk at the heads of the horses, the men behind with the plows. Pastures are scoured with ox-drawn dredges, plowlands broken up with horse-drawn harrows. The final cartloads of the previous summer's hay, which have been standing all winter in stacks out on the meadows, are being taken back to the barns before the cattle are let out onto the spring grazing. The only sound on the road is the oiled creak of the cart axles as they pass. No day is more wonderfully spent than in loading hay onto one of those carts, riding it home to the barn, and forking it up into its summer shelter.

Once seen, smelled, and heard, that thickly peopled world starts to make sense of the Wealden net of lanes and woods and narrow shaws, the strips of often coppiced trees that divide the fields. No machine was involved in the shaping of this. All tools were hand tools, all work handwork. Only the mills – like the ancient creaking structure at Bateman's that still grinds corn with the power of the Dudwell that is led to it in long shallow leats – were not driven by muscle power, animal or human.

The key ingredient was handwork. The Weald, whose beauty consists in its having been handmade, still bears the memory in its bones as a place of deep poverty and hard labor. Try digging a posthole in the clay or even burying a pet or a dead lamb in the corner of a field and you soon know how intractable the land is: the deep Wealden lanes were mostly impassable all winter, or would have needed a team of oxen to drag a wagon through their clag, and hardly a day of sunshine passes before that stolid stickiness transmutes into an equally unaddressable concrete. No one who could have chosen to farm elsewhere would have opted for these difficulties.

Everyone was subject to the tyranny of work. One ancient parishioner in Burwash in the 1850s described her life as a girl at the end of the eighteenth century:

> When I was sixteen years old I was had out, like a cow, to the market, and any farmer who wanted a servant come and choosed one. [In the farm where she went to work at Wadhurst] I'd churning twice a week, and cheesing twice a week, and brewing twice a week, beside washing and baking; and six cows to milk every night and morning, and sometimes a dozen pigs to feed. There were four men lived in the house, and I'd all the

The Weald, whose beauty consists in its having been handmade, still bears the memory in its bones as a place of deep poverty and hard labor.

bilin' to do – the cabbage and the peas and the pork for their dinners, besides all the beds to make.

She was paid a shilling and sixpence a week and at times would begin work at four in the morning, ending at midnight.

Intractability and the need for intense, body-wracking work created this landscape of interlinked privacies, a part of the world honeycombed with human life, the shaws planted with hornbeam, for its strength and its ability to make the best of hot-burning charcoal; ash for the all-important lightness for the handles of rakes or scythes; oak for everlasting robustness, to make the thatched houses, barns, and byres of which each separate farmstead consisted. Historically, fields were tiny, often no more than one or two acres. And farms were equivalently small, often no more than twenty or thirty acres, with a few cows, some pigs, and chickens. Oxen were the draft animals of choice, but this was farming on the scale of gardening.

Some acres were sown with wheat or oats, but the two great living products of the Weald were timber and hay. Much as in Transylvania now, it was a world that needed wood for its structures – not only the buildings but the fences around gardens and cattle yards, the hayracks, the hen coops, the pigsties – and which relied for sustenance on grass. It was a world built on hay, that steady transfer of nutrients from summer meadows to winter byres, without which the whole system would have collapsed. If you can still find a farmstead which retains its old buildings, you are really looking at an interfolded cohabitation of people, cattle, and grass.

Perhaps that is why, even with modern machinery, the hay harvest in July still feels like the deepest connection with this old world. "Haying," as it is called, on the same principle as "wooding" or "lambing," is the climax of the year, the point around which everything else turns. It is a moment when a grass field delivers up its riches, at first lying sleek and glossy like hanks of hair on a barber's floor, then baled and carted home as ingots of summer goodness. You won't find Sussex farmers getting starry-eyed about this haying moment, but there is something magical about well-made hay. "Look at that," Fred Groombridge, an old Brightling farmer, would always say to me when trying to sell me a bale or two of his hay. "Put your nose in it, Adam. You can smell the summer sunshine in it, can't you?" Fred always looked at me with one eye closed. "Why's that?" I asked a friend. "Because he's thinking with the other one."

It is a story of intense locality and short horizons, this field being essentially distinct from that one, and a beautiful rich local language derived from the Saxons after whom Sussex is named to describe it: to get muddy in the Weald is to be slubbed up; to work out a problem involves stirring it about a bit; the twiggy lengths falling from a cut hedge are the brishings; a fence post is a spile; a sickle is a swop. I think most of those words are on their way out now, but even twenty-five years ago when I went haying with Fred and his friends, that was the language they used, as old as the woods, older than the hedges.

"The wooded, dim blue goodness of the Weald" Kipling called it in a memorable phrase, but that romantic sense of a lost world seen at a distance is probably more a reflection of Kipling's own state of mind when retiring to

> You won't find Sussex farmers getting starry-eyed about this haying moment, but there is something magical about well-made hay.

Bateman's from the world than what this place would have been like when fully alive. Luckily, the great historians of localism in the Weald, David and Barbara Martin, have unearthed an account of everybody who was living and working in Ninfield in 1702, household by household, occupation by occupation, family by family. It is an annotated poll tax return, and what it reveals is a tiny, complex, nearly self-sufficient world of farmers, all of them part-time, with other occupations adding to the family finances. At the bottom end of the

Gill Bustamante, *Poppies in a Sussex Meadow*, oil on canvas, 2014 social scale are landless laborers: a sheep-shearer, a "collier," meaning a man who made charcoal, some carpenters, and two butchers. Many of these poor men had nothing but a small garden in which to grow their vegetables and several of them lived entirely on their own. Nevertheless, Ninfield in 1702 had its own shoemaker, glover, weaver, tanner, and two tailors. This was a place that was not only fed and housed from its own land but clothed and shod from it.

What are we left with? Only a landscape full of hidden corners and the lingering suggestion of the life that created it.

Rising in the social scale, above those bottom strata in which most farms were under fifteen acres and some under five, came the skilled craftsmen – a wheelwright, a miller, a carpenter, and a "bullock leech"– the cattle vet. Finally, a cluster of larger-scale men, with above fifty acres each, and variously classed as gentlemen and yeomen, the richest of whom farmed more than 400 acres. He had three servants living in his handsome Queen Anne house at Lower Standard Hill (now known as Luxford House) but like nearly everyone else in the village had no adult children living with him. This was a hard-won life, even for the relatively rich, and grown-up children had to be pushed out into the world.

Nothing could be more obvious than that this world has changed. Even in the thirty years I have known it, the Weald has lost much of its local distinctiveness: its ways of life, of speaking and thinking, of making a living, the connections to local stockmarkets (most of which have gone) and local abattoirs (ditto) have all become generalized, less distinct, less Wealden.

Those changes feed through into the economy. Of nearly 50,000 jobs in Sussex, barely 600 are now involved with the land. That figure has fallen by 400 in the last ten years. The average income for a lowland grass farm in England is £9,400 a year (plus the subsidy, whose future is uncertain), which in itself is enough to explain why young men and women are not going into farming. None of us likes the idea of churning twice a week, cheesing twice a week, or brewing twice a week, making the hay by hand, coppicing the hornbeam for our winter warmth, feeding pigs, or milking cows morning and night. It would be ridiculous to hanker after the material poverty of the past. Even in Transylvania, the same processes of delocalization are underway.

And so what are we left with? Only a landscape full of hidden corners and the lingering suggestion of the life that created it, the orchid meadows that have survived by chance or because one farmer valued a flowery summer enough not to "improve" his grassland; the woods, as Kipling wrote, that "know everything and say nothing," the consolations of a place which you have to walk through to know it, to encounter its memories through the soles of your feet. ⮞

The Glory
of the Creatures

✦

Readings for the Seven Days of Creation

Adam Names the Animals, from the *Aberdeen Bestiary*, twelfth century

First Day

✦

Light

THE BOOK OF JOB

And the Lord answered Job from the whirlwind and He said:
Who is this who darkens counsel in words without knowledge?
Gird your loins like a man,
that I may ask you, and you can inform Me.
Where were you when I founded earth?
Tell, if you know understanding.
Who fixed its measures, do you know,
Or who stretched a line upon it?
In what were its sockets sunk,
or who laid its cornerstone,
when the morning stars sang together,
and all the sons of God shouted for joy?

Robert Alter, trans., *The Hebrew Bible* (W. W. Norton & Company, 2019), 563–564.

The artwork accompanying these readings is taken from an account of the Creation found in the Nuremberg Chronicle, *a universal history compiled by the Nuremberg doctor, humanist, and bibliophile Hartmann Schedel (1440–1514). Completed in 1493, it is one of the earliest printed books to successfully integrate illustrations and text.*

Second Day

•

Firmament

BLAISE PASCAL (1623–1662)

Let man then contemplate the whole of nature in her full and grand majesty, and turn his vision from the low objects which surround him. Let him gaze on that brilliant light, set like an eternal lamp to illumine the universe; let the earth appear to him a point in comparison with the vast circle described by the sun; and let him wonder at the fact that this vast circle is itself but a very fine point in comparison with that described by the stars in their revolution round the firmament. But if our view be arrested there, let our imagination pass beyond; it will sooner exhaust the power of conception than nature that of supplying material for conception. The whole visible world is only an imperceptible atom in the ample bosom of nature. No idea approaches it. We may enlarge our conceptions beyond all imaginable space; we only produce atoms in comparison with the reality of things. It is an infinite sphere, the center of which is everywhere, the circumference nowhere. In short it is the greatest sensible mark of the almighty power of God, that imagination loses itself in that thought.

Blaise Pascal, *Pascal's Pensées*, trans. W. F. Trotter (E. P. Dutton, 1958), 17.

Third Day

＋

Trees & Fruit

JULIAN OF NORWICH (1342–ca. 1416)

At the same time, our Lord showed me a spiritual vision of his familiar love. . . . In this vision he also showed a little thing, the size of a hazelnut in the palm of my hand, and it was as round as a ball. I looked at it with my mind's eye and thought, "What can this be?" And the answer came to me, "It is all that is made." I wondered how it could last, for it was so small I thought it might suddenly have disappeared. And the answer in my mind was, "It lasts and will last for ever because God loves it; and everything exists in the same way by the love of God." In this little thing I saw three properties: the first is that God made it, the second is that God loves it, the third is that God cares for it. But what the maker, the carer and the lover really is to me, I cannot tell; for until I become one substance with him, I can never have complete rest or true happiness; that is to say, until I am so bound to him that there is no created thing between my God and me. We need to know the littleness of all created beings and to set at nothing everything that is made in order to love and possess God who is unmade.

Julian of Norwich, *Revelations of Divine Love*, ed. Clifton Wolters (Penguin Books, 1966).

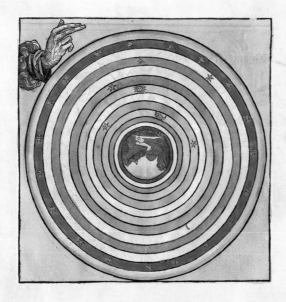

Fourth Day

◆

Sun, Moon & Stars

FRANCIS OF ASSISI (1181–1226)

Most High, all-powerful, good Lord.
Yours are the praises, the glory, the honor, and all blessing.
To You alone, Most High, do they belong,
and no man is worthy to mention Your name.
Praised be You, my Lord, with all your creatures,
especially Sir Brother Sun,
Who is the day and through whom You give us light.
And he is beautiful and radiant with great splendor;
and bears a likeness of You, Most High One.
Praised be You, my Lord, through Sister Moon and the stars,
in heaven You formed them clear and precious and beautiful. . . .
Praised be You, my Lord, through Sister Water,
which is very useful and humble and precious and chaste.
Praised be You, my Lord, through Brother Fire,
through whom You light the night
and he is beautiful and playful and robust and strong. . . .
Praise and bless my Lord and give Him thanks
and serve Him with great humility.

Francis of Assisi, *Francis and Clare: The Complete Works*, trans. Regis J. Armstrong and
Ignatius C. Brady (Paulist Press, 1982), 38–39.

Fifth Day

◆

Birds

MECHTHILD OF MAGDEBURG (ca. 1207–ca. 1282)

If a bird remains on the ground for a long time, its wings deteriorate and its feathers become heavy. Then it raises itself aloft beating its feathers and drawing itself far upward until it catches the air. Then it takes flight yet higher. The longer it flies, the more it soars in exhilaration, scarcely returning to earth to refresh itself. Thus did the wings of love take from it earthly pleasure. In this same way we should prepare ourselves when we are to approach God. We should raise the feathers of our longing to God. We should elevate our virtue and our good works with love. If we do not give up on this, we shall become conscious of God.

Mechthild of Magdeburg, *The Flowing Light of the Godhead*, trans. Frank Tobin (Paulist Press, 1998), 329.

Creation of the Animals, *Aberdeen Bestiary* (detail)

Sixth Day

⬥

Animals

CHRISTOPHER SMART (1722–1771)

For I will consider my Cat Jeoffry.

For he is the servant of the Living God duly and daily serving him.

For at the first glance of the glory of God in the East he worships in his way.

For this is done by wreathing his body seven times round with elegant quickness.

For then he leaps up to catch the musk, which is the blessing of God upon his prayer.

For he rolls upon prank to work it in. . . .

For if he meets another cat he will kiss her in kindness.

For when he takes his prey he plays with it to give it a chance.

For one mouse in seven escapes by his dallying.

For when his day's work is done his business more properly begins.

For he keeps the Lord's watch in the night against the adversary.

For he counteracts the powers of darkness by his electrical skin and glaring eyes.

For he counteracts the Devil, who is death, by brisking about the life.

For in his morning orisons he loves the sun and the sun loves him.

For he is of the tribe of Tiger. . . .

For he is a mixture of gravity and waggery.

For he knows that God is his Saviour.

Christopher Smart, *Rejoice in the Lamb: A Song from Bedlam*, ed. William Force Stead (J. Cape, 1939).

Sixth Day

◆

Man

AUGUSTINE OF HIPPO (354–430)

I would not exist, my God, I would not exist at all,
unless you existed in me. Or is it rather that I would
not exist unless I existed in you, from whom, through
whom, in whom, everything exists? That's it, Master,
that's it. To what place can I call you, if I am in you?
And from what place can you come into me? Where
would it be, outside heaven and earth, that I could
withdraw, so that God could come into me there – the
God who said, "Heaven and earth are filled with me"?

Augustine, *Confessions*, trans. Sarah Ruden (Modern Library, 2018), 5.

Seventh Day

•

And It Was Good

SADHU SUNDAR SINGH (1889–1929)

God is revealed in the book of nature for God is its author.
Yet we only comprehend this book if we have the necessary
spiritual insight. Without reverence and perception we
go astray. We cannot judge the truthfulness of any book
merely by reading it. Agnostics and skeptics, for example,
find only defects instead of perfection. Skeptics ask, "If
there is an almighty creator, why then are there hurricanes,
earthquakes, pain, suffering, death, etc.?" This is like
criticizing an unfinished building or incomplete painting.
When we see them fully finished, we are embarrassed at
our own folly and praise the skill of the artist. God did
not shape the world into its present form in a single day,
nor will it be perfected in a single day. The whole creation
moves toward completion, and if we see it with the eyes
of God moving toward the perfect world without fault or
blemish, then we can only bow humbly before our creator
and exclaim, "It is very good." ⤳

Sadhu Sundar Singh, *Wisdom of the Sadhu* (Plough, 2014), 57–58.

Regenerating the Land

An Amish Farmer's Quest to Heal Agriculture

AN INTERVIEW WITH JOHN KEMPF

Regenerative agricultural practices seek to improve plant health by restoring and maintaining natural soil biology and chemistry. Ohio Amish farmer John Kempf has emerged as a leading advocate, with a podcast, webinars, and a consulting firm, Advancing Eco Agriculture. Jeff King, who applies these principles to grow vegetables for his Bruderhof community in New York, interviewed him for Plough.

Jeff King: Is regenerative agriculture just a new name for organic farming?

John Kempf: There are people seeking to come up with regenerative certifications; I hope they remain unsuccessful. Organic certification unfortunately devolved into an us-versus-them conversation where one side is good and the other is evil. I'm really glad that the conversation around regeneration still has an openness and the recognition that this is a journey, not a destination. A farmer just starting out, who has been engaged in practices that were very degrading to the land because he didn't know any different, when he becomes aware, can begin making that transition. We should embrace that grower, welcome him, and give him the support and encouragement he needs.

You talk about the moment at which growers realize that the track they're on is a disastrous one. Did you have that kind of experience yourself?

Yeah, I had a two-by-four-in-the-face moment, on the family fruit and vegetable farm I grew up on, about fifteen acres in the snowbelt south of Lake Erie. We would get anywhere from ninety to a hundred inches of snowfall per year, and total rainfall of thirty-eight to forty-two inches per year. I used to believe this wet environment gave our crops a predisposition to being susceptible to diseases and insects.

My dad started growing vegetables in 1996. When I finished school at fourteen, I was given the responsibility of doing all the drip irrigation and the foliar applications of fertilizers and biostimulants and pesticides. My dad was also the regional pesticide supplier; we were the first people to try all the newest pesticides and then make recommendations to farmers based on how well they were working for us. But in spite of ever-intensifying pesticide use, it became more and more difficult to control different diseases and insects. We had a three-year period, 2002 to 2004, in which we lost upwards of 70 percent of our four primary crops.

> **When we grow crops that are resistant to diseases, they also improve the health of the people and the livestock who consume these plants as food.**

In the 2004 growing season, we rented a field from a neighboring dairy farmer that bordered one of our own fields. These two fields used to be tilled and planted up and down the slope but we switched the direction of tillage and planted the crops across the field border. On one end, we planted a crop of cantaloupe.

At harvest time, on the soil we had been farming for the prior decade with intense pesticide applications, 80 percent of the leaves were infected with powdery mildew. On the new soil, from the dairy farm with a four-year rotation of corn, small grains, and two years of hay without pesticide applications, there was no powdery mildew. Not 5 percent infection or 10 percent infection, but zero – you couldn't find any powdery mildew in that part of the field; there was a knife-line effect right down the center. The vines were the same variety, planted the same day, same fertilizer applications, same fungicide applications, but two completely different outcomes.

I wanted to know: What are the differences? What allows one plant to be resistant to powdery mildew when the plant two feet away is susceptible? I was very fortunate to connect with several amazing mentors, and I learned a lot about plant physiology and plant immune systems – plants have an immune system much the same as we do. There's a significant body of research describing how to manage plant nutrition to increase their immune systems and develop their resistance to diseases and insects. But what really got me started down the pathway of regenerative agriculture was the realization that when we grow crops that are resistant to diseases, they also improve the health of the people and the livestock who consume these plants as food.

You've claimed that healthy plants can be completely resistant to any disease or insect. That's a pretty audacious claim.

This is not a theory. This is something that we have actually put into practice on millions of acres here in North America, very successfully. It is possible for crops to be resistant to grasshoppers, to aphids, to Colorado potato beetles, to Japanese beetles. There are obviously hundreds of diseases and insects and we haven't worked with all of them, but we have yet to encounter one we have not been successful in treating and reversing with nutrition management, including a number of diseases for which there is no known pesticide treatment.

Can you give an example of a specific farmer who has been able to do this?

Let's talk about one of the incurable diseases. Bacterial canker on cherries is an infection where the bacteria live inside the plant's vascular tissue, and there is no known pesticide treatment. Back in 2011, one of the preeminent cherry growers in the Pacific Northwest, said to us, "I don't care about reducing pesticides. I

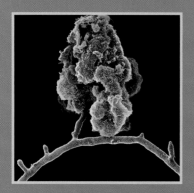
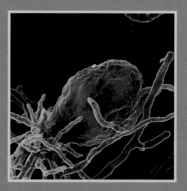
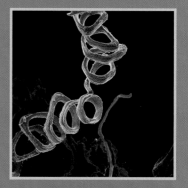

Healthy soil hosts a beneficial community of bacteria, fungi, and other microscopic organisms.

Each spoonful of soil contains up to one billion of these microorganisms, which are essential to plant growth.

Scientists are only beginning to understand the complex relationships within this microbiome.

Helium ion microscopy by Shuttha Shutthanandan, Pacific Northwest National Laboratory.

have no desire to reduce my fertilizer applications. And I do not want to be organic. What I really want is to grow large, firm cherries that qualify for the export market to Asia. And I've heard that you can help me do that."

He gave us fifty acres of cherries to run trials on, a block of trees about five years old and so badly diseased that, if we weren't able to turn it around in one year, they were going to take the bulldozer to it and replant. Eighteen months after we started treating the diseased trees there was no detectable bacterial canker anywhere in that block.

One of our first recommendations was that he was putting on too much fertilizer too early in the season, and we tested it. In one block of trees, half got the fertilizer, half did not. At the end of harvest season, he concluded that stopping that fertilizer application improved his profitability by $1,600 per acre. This has been one of our foundational realizations: the majority of disease and insect problems are not caused by nutrient deficiencies but by the excesses of products farmers apply.

Some people think regenerative agriculture might be less productive, but our experience has been the opposite. When you manage plant nutrition for optimal quality and health, yields increase. Our message to farmers is very simple: we can help you be more profitable by managing plant nutrition differently.

You've hinted at the concept of food as medicine, which is a very powerful thought for growers like me who want to be effective in helping people and have the biggest positive impact we can.

We have a public health crisis around the world today. We have all these degenerative illnesses such as heart disease and diabetes and stroke and cancer that, when you combine them, afflict the majority of our population. Fundamentally, I would suggest that this is a failure of agriculture to produce nutritious food that can enhance our immune systems. There's certainly more involved than just the food we eat, but food certainly is a significant contributing factor.

All plants produce compounds generically referred to as plant-secondary metabolites. Some of these compounds whose human health benefits have been relatively well-studied would be lycopene in tomatoes, or resveratrol in red wine and watermelons, or

Photograph by Darius Clement. Used by permission.

Cover crops such as the winter rye in this canteloupe field form a key part of the regenerative approach.

anthocyanins in blueberries. The piece that has been missed is that the concentrations of these compounds within plants vary dramatically based on the plant's health. In 1992, Michigan State University conducted research and found that the anthocyanin content of blueberries can vary by a factor of twenty-five.

We've also forgotten what really good food tastes like. Can you imagine the flavor and the aroma differences between blueberries that have twenty-five times variability in anthocyanin content? The same is true if you're speaking about tomatoes or spinach or strawberries. Even when we grow our own fruits and vegetables in our gardens, unless we address nutrition management and specifically seek to grow nutrient-dense, high-quality fruits and vegetables, most of us have no idea what really healthy fruit and vegetables actually taste like.

When I watched one of your webinars for Midwestern corn and soy growers, I was surprised to hear you quoting Genesis and Job.

Modern farming adopted a model of agriculture that is very degrading of our ecosystems, one that has contributed tremendously to a loss of topsoil, a loss of nutritious food, a loss of water quality. You think about that from a Christian perspective and you ask, "How do we profess to be stewards of God's creation and yet have adopted a model of agriculture with such a negative impact on the creation that we are here to steward?"

I came to the conclusion that this revolves around two incorrect beliefs. The first is that we are here to have dominion over the earth, a common translation of the Hebrew in Genesis. In modern English, "dominion" implies domination and subjugation. But the original intent is clearly to communicate that we have a responsibility to steward God's creation.

The second belief is that the earth is cursed. We don't even have to go into the New Testament and the act of salvation to refute this. We simply have to look at Genesis 8:21. After Noah emerged from the ark, God smelled his burnt offering and said, "Never again will I curse the ground because of humans, even though every inclination of the human heart is evil from childhood." So if we embrace the promises God has given us, we can adopt a model of agriculture that avoids pestilences and plagues, diseases and insects and weeds. Ultimately, if we believe that the land is cursed, then it will be – but it doesn't have to be.

We see the language of warfare being applied in agriculture, even "nuking" pests or weeds. So yours is perhaps a more nonviolent approach to farming?

You're absolutely right. We have chemicals called Panther and Prowl and Warrior, based on this mentality of search and destroy – if the first weapon of choice is not successful, simply get a bigger bomb. This is not an ethos of stewardship. Job 5:23 comes to mind: "For you shall have a covenant with the stones of the field, and the beasts of the field shall be at peace with you." We should work in alignment with the minerals and the nutrients that give us abundant plant nutrition and quality, because that is what gives resistance to drought and the resilience to continue to produce crops in very challenged environments. A covenant is not a one-sided relationship. And farmers want to do the right thing. No farmer has ever got out of bed in the morning and thought, "What can I destroy today?"

It seems a lot of these farmers are specializing in one crop. Should we be moving away from this kind of monoculture and toward more diversity in our planting? Might the relationship between crops and livestock be part of that?

I do believe it's necessary for us to incorporate livestock back into the landscape. That step can be more effective at regenerating ecosystems than almost anything else. We also need to

develop regionalized and decentralized food and agricultural systems. I don't think it's healthy from a national food-security perspective that 45 percent of our entire fruit and vegetable supply in North America is produced in three counties in California. We once had vibrant fruit and vegetable production throughout the Midwest. The Great Plains have some of the most fertile soils on the globe. Are we really going to use those incredible, beautiful, fertile soils to produce ethanol?

> **How do we profess to be stewards of God's creation and yet have adopted a model of agriculture with such a negative impact on the creation that we are here to steward?**

Would you agree that a farm's fertility should be maintained from within the farm's borders?

In concept, yes. There should be fewer fertilizer and pesticide inputs from such external sources as mined fertilizers. But what if your operation is a one-and-a-half-acre market garden? You don't have room for livestock. The resources to maintain and regenerate soil fertility should still come from the local ecosystem, but that can be from a local livestock farmer producing compost.

Another principle of regenerative agriculture is reducing tillage. For growers I know, that's one of the biggest hurdles. Is it necessary to go completely no-till?

I dislike the way this conversation has been framed – it's these binary, polar opposites, good and bad, black and white, and it's not that simple. Indigenous populations have

would be no tomatoes that tasted like cardboard and no strawberries that were picked half-green! What determines what arrives at the grocery store is the processors and distributors. Farmers also do not desire to grow cantaloupe that tastes like a rock. They enjoy growing flavorful food.

Have you experienced opposition from corporations with a vested interest in the status quo?

Actually, we now have CEOs of large ag-chemical corporations talking about how they will be helping farmers adopt regenerative agriculture practices. It's debatable how real that is, but at least they recognize that this movement is too significant to squash.

I see myself as providing information to the people who really hold the power to change the way agriculture is done – the farmers themselves. Farmers can make the choice to discontinue pesticide and fertilizer applications, and if they do, suppliers are largely powerless to change them.

It's not beneficial to be anti-Monsanto, anti-GMO, anti-Roundup, because the moment you take a position, you immediately alienate all the people who are currently using those tools and technologies. That limits our ability to facilitate a transition. When we talk about having empathy with plants and livestock and ecosystems and the land, that also extends to having empathy with people who have a different perspective, who might be the "them" in "us-versus-them." The more successful we are at approaching dialogue from a place of love and understanding, the more successful we will be at creating systemic change. ⇒

Interview conducted February 4, 2021.

Regenerative methods work even for conventionally high-spray crops such as stone fruit.

been growing potatoes for thousands of years in the Andes; those soils have been tilled continuously and they are not degrading. We have to look at the net balance, not just one single factor. So if tilling might have a negative impact, then what are we doing to counterbalance that impact? Are we growing cover crops? Are we adding compost and microbial inoculants and biostimulants?

That being said, I am an advocate of reducing or eliminating tillage. There has been incredibly successful innovation on smaller farms, but it doesn't always transfer well to fifty acres or five hundred acres. We desire to develop an agriculture over time that is more regionalized, more decentralized, more scale-appropriate. In the meantime, we have to recognize the reality that exists today, where seven thousand fruit and vegetable growers in the United States produce 90 percent of the total domestic supply chain – we need to develop regenerative systems that can be executed successfully on thirty thousand acres of carrots.

You're saying what will drive change is economic interest. What about consumer demand?

Consumer demand does not determine what arrives at the supermarket shelf. If it did, there

Len Stomski, *Spotlight*,
oil on gessoed panel

The Path

It's not a path that takes you very far.
It starts across the field from where you're standing
but only brings you back to where you are.

You try it, like a door that's left ajar.
A little uphill climb, not too demanding;
it's not a path that takes you very far.

It's better, though, than sitting in the car.
The view, while not what you would call "commanding,"
at least gives you a sense of where you are.

Between two pines you glimpse the reservoir
where swallows briefly rest before disbanding.
Their paths, like yours, don't take them very far.

The catbird practicing its repertoire,
the squirrel perched above you, reprimanding,
return you to the sense that, where you are—

though not a garden painted by Renoir—
the monarch's ever on the verge of landing.
It's not a path that takes you very far.
It only brings you back to where you are.

ALFRED NICOL

Let the Body Testify

Whose Bodies Matter?

LEAH LIBRESCO SARGEANT

UNTIL THE 1980S, newborn babies who needed surgery went under the knife with no pain medication or anesthesia.[1] Doctors assumed that they were too young to feel pain – or at least to feel pain the way an adult does, which, for the professionals, was the only way that counted. One 1943 study helped shape this assumption.[2] Myrtle McGraw had pricked resting, swaddled infants with a pin and judged the children's response to be minimal.[3] She had pricked them, and they did not bleed.

When the babies cried and screamed as a surgeon cut into them, their reaction was dismissed as reflex, not a real experience. The babies could speak for themselves at great volume, but not in a way their doctors were willing or able to listen to.

The vulnerability of our bodies is part of what binds us together into a community. In Jesus' parable of the Good Samaritan, the story begins with the traveler's suffering when he is beaten and robbed. His need is what calls neighborliness out of the Good Samaritan, who binds the traveler's wounds, takes him to a refuge, and ensures his continued care.

This story is Christ's answer to an expert in the law, who asks Jesus to clarify the limits of the Great Commandment. God calls me to love my neighbor as myself, but who, exactly, counts as my neighbor? And, left as the subtext, who *doesn't* count? Whom am I allowed to *not* love?

The surgeons entrusted with tiny, vulnerable patients managed to pass by the babies' need. They hurried about their lifesaving business without seeing *people* in their patients. There should be nothing more unignorable than a baby's shriek, but, when we don't believe in the dignity of the person, we find ways of denying her body and her pain.

In abortion facilities, the bodies of babies are painstakingly reassembled in "products of conception" rooms. The doctors must verify that the child is whole, before the body is thrown away as medical waste. A single forgotten limb left in the mother's womb is an invitation to sepsis and rot.[4] A body ignored is corrosive.

Our bodies are a brute fact. The immediacy of a sprained ankle is an aching interruption to our routines, our sleep, our thoughts. Chronic fatigue poses a constant question: If I do this, how will I pay for it later? But to get help from others, we have to find a way to cry out so that our neighbor will hear us.

Babies are too young to be able to match the expectations of the people who care for them. They are starkly honest. But, as we grow up, only some bodies will be heard and recognized by people in power. Women, people of color, the disabled all find themselves needing to translate their experience in order to be heard.

Shai Yossef, *Newborn,* oil on canvas

1. Rachel Rabkin Peachman, "Why Aren't We Managing Children's Pain?", *New York Times,* June 27, 2016.
2. George Dvorsky, "Why Are So Many Newborns Still Being Denied Pain Relief?", *Gizmodo,* February 4, 2016.
3. Myrtle B. McGraw, "The Neuromuscular Maturation of the Human Infant," *Psychosomatic Medicine,* January 1944.
4. Stephanie Armour, "Two Women Spotlight Two Sides of Abortion Debate," *Wall Street Journal,* April 13, 2018.

Leah Libresco Sargeant, a Plough *contributing editor, is the author of* Arriving at Amen *(Ave Maria Press, 2015) and* Building the Benedict Option: A Guide to Gathering Two or Three Together in His Name *(Ignatius Press, 2018).*

Making our internal experience externally legible may mean leaving out details, playing up to stereotypes, or otherwise matching what our neighbor *expects* to hear, whether or not it matches what we need to say.

Legibility, in this sense, is a concept popularized by James C. Scott in his 1998 book *Seeing Like a State*. He describes legibility as a central problem in statecraft – the larger the state, the more effort it must put into being able to standardize its people so that they can be "seen" by the state apparatus.[5] Legibility is why states assign last names to people who previously lacked them or addresses to locations that were described solely by reference to local landmarks.

> We are made in the image of God, and some part of him is denied when the goodness of the people he has made is denied.

Scott is suspicious of projects to render people and places legible, finding that they often oversimplify and flatten natural relationships. A planned, gridded forest may suffer soil collapse due to the lack of complementary plants which were treated as irrelevant weeds. Scott recommends cultivating a degree of illegibility, in order to remain more independent of state programs and oversight.

Unchosen illegibility, however, means being overlooked. Women are not more free because they are less thought of. In *Invisible Women: Exposing Data Bias in a World Designed for Men* (2019), Caroline Criado-Perez assembles a long list of places women's bodies and women's needs are ignored. Bricks are sized to be easily held and handled by a man's larger hand.[6] Drug dosages are calibrated to men's larger bodies, leaving women overmedicated and struggling with side effects.[7] Even car voice assistants (with their deferential female speech patterns) are tuned to hear male voices. Criado-Perez coaches her own mother to lower her voice into a male register in order to be heard by her car's computer.[8]

When confronted with this problem, one auto parts executive suggested that "many issues with women's voices could be fixed if female drivers were willing to sit through lengthy training," as *Autoblog* put it in 2011.[9] In this view, the onus is on women to change for the technology, rather than the other way around.

Altering your voice might seem a trivial example, but this is the least of the ways society pressures women to alter themselves to meet others' expectations. From unrealistic beauty norms to unnatural footwear, the ways in which women are asked to change their bodies to fit expectations can seem unending. For those with the time, money, and genes to pursue it successfully, this strategy can be effective, but is also exhausting. It requires us to be untruthful, or at least only partially truthful about who we are. We are made in the image of God, and some part of him is denied when the goodness of the people he has made is denied.

Suffering Correctly

In his provocative 2010 book, *Crazy Like Us: The Globalization of the American Psyche*, Ethan Watters argues that some mental illnesses take their shape from the expectations of the culture. Anorexia, schizophrenia,

5. James C. Scott, *Seeing Like a State: How Certain Schemes to Improve the Human Condition Have Failed* (Yale University Press, 1998).

6. Caroline Criado-Perez, "The Deadly Truth about a World Built for Men," *Guardian*, February 23, 2019.
7. Roni Caryn Rabin, "The Drug Dose Gender Gap," *New York Times*, January 28, 2013.
8. Caroline Criado-Perez, *Invisible Women: Exposing Data Bias in a World Designed for Men* (Harry N. Abrams, 2019).
9. Sharon Silke Carty, "Many Cars Tone Deaf to Women's Voices," *Autoblog*, May 31, 2011.

and post-traumatic stress disorder are more like grief than like a physical ailment such as gangrene. All people grieve, but the form that mourning takes varies by culture. In the United States, we often wear black to a funeral, while in China, the funereal color is white. Jews traditionally rend their garments while sitting shiva, while Victorians had a half-life to their mourning clothes, fading from black to grey to mauve. The clothes make our sorrow legible to our community.

Watters argues that certain disorders are a way of giving voice to anguish in the language it will be heard in. His aim is not to explain away mental illness – the disorder gives voice to something real – but he believes we teach each other how to suffer, just as a community creates norms around mourning. And in an increasingly globalized world, America has, in the guise of aid, been homogenizing the world's experience of pain, flattening foreign bodies to make them legible to our doctors.

In one of his examples, American medical expectations touch off an anorexia epidemic in Hong Kong. A well-publicized case of a fourteen-year-old who starved herself to death in 1994 prompted a comprehensive program to raise awareness of the disease, and succeeded too well. Dr. Sing Lee, a specialist in eating disorders, had previously seen two to three anorexic patients a year, but, after the publicity blitz, he began receiving that many referrals a week.

His patients' experience of their disease had changed as its prevalence increased. Initially, the rare anorexics he saw didn't know there was a name for their condition. They told him that they couldn't eat, not that they feared being fat. They could accurately describe and draw their bodies, instead of holding on to a

Shai Yossef, *Sisters,* oil on canvas

Shai Yossef, *Love,* oil on canvas

distorted self-image. But as the disease was publicized, the women he saw fit the *Diagnostic and Statistical Manual of Mental Disorders'* criteria more and more closely.[10]

It's no coincidence Watters saw this dynamic play out in anorexia, which is most prevalent among women. Women are particularly vulnerable to pressure to make their ailments conform to expectations. Watters cites Canadian scholar Edward Shorter's theory for this pressure to conform. As Watters explains it, "People at a given moment in history in need of expressing their psychological suffering have a limited number of symptoms to choose from – a 'symptom pool.'"[11] Without actively intending it, people in distress rely on our expectations of illness to find a way of being recognized.

My own middle-school health classes veered close to this approach. With the best of intentions, the teachers gravely instructed us that many girls experienced disordered eating, that high-achieving girls might be drawn to calorie-counting as one more thing to excel at, that it could be an exhilarating way of experiencing control if you lacked it elsewhere. It became a tutorial in how to suffer *correctly.*

Today, a girl who experiences trauma or distress as her body changes, her desires stir, and men sexualize and harass her is more likely to be told that one option from the symptom pool is to not be a woman. In those same health classes, she may be told that discomfort with her changing body is evidence she might belong in a different body.[12] In *Prospect,* Emma Hartley considers the flip in gender dysphoria referrals from 75 percent male to

10. Sing Lee, "Anorexia Nervosa in Hong Kong: A Chinese Perspective," *Psychological Medicine,* August 1991.
11. Ethan Watters, *Crazy Like Us: The Globalization of the American Psyche* (Simon & Schuster, 2010), 32.
12. "Principles of Gender Inclusive Puberty and Health Education," Gender Spectrum 2019 report.

70 percent female in less than a decade and explores what pressures and prejudices may contribute to this trend: "This is a story that needs to be understood at the level of society, not just the individual psyche."[13]

This theory of gender is just an update and expansion of what women have been hearing for years – that our bodies and our selves are the problem; that when there is a mismatch between us and our society's expectations, it is we who have to change.

Women routinely alter their bodies to fit these expectations. They dye their hair and receive cosmetic injections in order to avoid the appearance of aging. Women take contraceptive pills to avoid the natural hormonal cycle, and the risk of pregnancy, instead existing in a chemical state corresponding to perpetual pregnancy (minus the baby). Over one in five American women over forty report that they took antidepressants in the last thirty days, more than double the rate of American men.[14] To fit the space allowed, women change their appearance, their bodies, their feelings.

Escaping Expressive Individualism

There is one further demand placed on women and others who don't fit the "standard" body. They are asked to accept this burden of reshaping themselves as an opportunity for empowerment and self-expression.

In *What It Means to Be Human: The Case for the Body in Public Bioethics* (2020), O. Carter Snead traces changes in how we view the body, particularly in bioethics. He sees a conflict between two claims about the source of human dignity: is our worth rooted in our existence as living bodies, or as disembodied wills?

If we are primarily wills, the human person is valuable because of his or her ability to *choose*. Snead refers to this ideology as "expressive individualism." Former Supreme Court Justice Anthony Kennedy's decision in *Planned Parenthood v. Casey* was rooted in this understanding of human worth. In Kennedy's sweeping declaration, "At the heart of liberty is the right to define one's own concept of existence, of meaning, of the universe, and of the mystery of human life."[15]

This perspective promotes a kind of navel-gazing. "Flourishing is achieved by turning inward to interrogate the self's own deepest sentiments to discern the wholly unique and original truths about its purpose and destiny," as Snead puts it. "The truth about the self is thus not determined externally, and sometimes must be pursued counter-culturally, over and above the mores of one's community."

The model of expressive individualism sanctifies nearly any choice. In this framework, abortion is liberation for both mother and child. The Planned Parenthood slogan "Every child a wanted child" confers a peculiar dignity on survivors of a pro-choice world – unlike their aborted brothers and sisters, they were *chosen*.

This emphasis on choice puts a heavy burden on pregnant women. The unplanned, unasked-for child is cast as a failure, along with his or her mother. In an essay for *The*

> **Without actively intending it, people in distress rely on our expectations of illness to find a way of being recognized.**

13. Emma Hartley, "Why Do So Many Teenage Girls Want to Change Gender?," *Prospect*, March 3, 2020.

14. Debra J. Brody and Quiping Gu, "Antidepressant Use Among Adults: United States, 2015-2018," NCHS Data Brief No. 377, September 2020.

15. Anthony M. Kennedy, *Planned Parenthood v. Casey* majority opinion, 1992.

Atlantic, Sarah Zhang counted the cost of Denmark's universal screening for Down syndrome. Having more information made some families feel worse than if they were being kept in the dark.

The introduction of a choice reshapes the terrain on which we all stand. To opt out of testing is to become someone who *chose* to opt out. To test and end a pregnancy because of Down syndrome is to become someone who *chose* not to have a child with a disability. To test and continue the pregnancy after a Down syndrome diagnosis is to become someone who *chose* to have a child with a disability. Each choice puts you behind one demarcating line or another.[16]

> Every body is a testimony: we are made in God's image.

When their child was a choice, for parents to have or abort a child with a disability became part of their expressive identity. If their child had been disabled after birth in an accident, the parents would not feel that everyone who knew them looked at their child's condition as a choice that the parents had made. It would be simply who their child was.

Snead sees an alternative to expressive individualism – valuing the givenness of our bodies as they are, in all their vulnerability and weakness. In our fragility, he sees proof that we are relational beings.

> Because human beings live and negotiate the world *as bodies*, they are necessarily subject to vulnerability, dependence, and finitude common to all living embodied beings, with all of the attendant challenges and gifts that follow. . . . Given the way human beings come into the world,

from the beginning they depend on the beneficence and support of others for their very lives.[17]

Women are bound more tightly to this truth, even when society asks them to deny it. Not every woman will bear a child, but every woman lives with an awareness of her potential for new life, whether she experiences it as a gift or a threat. Even a planned, chosen pregnancy is a tutorial in the limits of the will.

It seems that won't stop people from using every technological advance at their disposal to bypass those limits nature would impose. In the world of assisted reproduction, meanwhile, the primacy of will radically reconfigures these embodied relationships, parceling out roles – egg donor, gestational surrogate, etc. – that multiply parents while muddying their connection and duties to the child.

Surrogacy contracts assert the rights of the parents to kill their child, even over the objections of the woman who is, moment by moment, sustaining that child. Parents might demand abortion as a resolution to a higher risk pregnancy of twins or triplets, or upon learning that their child carries a congenital abnormality. Their contracts claim that they have the higher claim on their child than the unrelated woman whose body has remade itself to help the baby grow, who to them is merely a commodity. To be a parent, in this understanding, is to have the authority to destroy.

In one case, when a surrogate named Melissa Cook offered to adopt the unwanted triplet rather than allow it to be killed, the father objected, out of a sense of justice. As Katie O'Reilly reported for *The Atlantic*, "The father of Melissa Cook's fetuses has stated that he believes singling one child out for adoption would be cruel, and thus he prefers to

16. Sarah Zhang, "The Last Children of Down Syndrome," *Atlantic*, December 2020.

17. O. Carter Snead, *What It Means to Be Human: The Case for the Body in Public Bioethics* (Harvard University Press, 2020), 88.

reduce."[18] He did not want to let one child live unchosen – judging it better to not be.

In such cases, a child who does not match our expectations or plans for our expressive identity may be deprived of life. After birth, people who don't fit in neatly may not face death but may still be negated, forced to be less themselves to be "allowed" to take up space. And, in countries with expansive euthanasia regimes, some people keep hearing the suggestion that suicide is the solution to the problem of their presence.[19]

18. Katie O'Reilly, "When Parents and Surrogates Disagree on Abortion," *Atlantic*, February 18, 2016.
19. Rachel Aviv, "The Death Treatment," *New Yorker*, June 22, 2015.

The narrower our ideas about whose bodies matter – who is our neighbor – the less likely we will be to help, love, or even see others. And with the body unacknowledged, it is easier to overlook the more permanent but more elusive soul.

Every body is a testimony: we are made in God's image. Our frailties reflect his commandment that we must love each other as he has loved us. When we marginalize our neighbors, we blot out that image and refuse the duty of that commandment. A woman's body, accordingly, does not need to be rewritten – women must be seen and loved as women. When we fail each other in this duty of love, our neighbor's body testifies against us.

Shai Yossef, *Touchdown*, oil on canvas

More Fish than Sauce

Beneath Panama City's gleaming
skyscrapers, traditional fishermen still
venture out to sea for a hard-won catch.

IVÁN BERNAL MARÍN

MIGUEL RODRÍGUEZ URIOLA cannot help swaying as he talks. He has spent half of his sixty-four years at sea, and it is as if the earth were moving beneath him like water. He was fourteen when he made his first voyage. Back then, the only way out of the city by boat was through a narrow tidal waterway, rowing against the muddy current with all the strength his arms could muster. Now his boat, like the other thirty-nine vessels registered in Boca la Caja, is powered by motors.

He has just returned from a six-day job out on the Pacific. He's sitting down by the entrance to the neighborhood, drinking a beer, the alcohol from the previous four already in his system. An outboard engine with clapped-out propellers hangs nearby, along with the shell of an old air-conditioning unit whose cooling days are long past.

Without letting the bottle fall, he gestures, as if throwing an imaginary trammel net, and says: "All this used to be wooden huts."

Past a vegetable-oil factory, surrounded by buildings which shoot up into the clouds, one of the last fishing neighborhoods in Panama City is surviving. The majority of its inhabitants, like Miguel, make their living by fishing. They are the last fully urban fishermen in Panama. Today the only indication of the surviving trade is a cardboard sign with shaky handwriting that reads: "We have prawns: 3.50 dollar."

Cars do not fit on Boca's streets. It lies at sea level and below traffic level. There are, however, enough antennas to ensure that the television signal has no problem getting through the surrounding wall of skyscrapers.

Boca's heart beats more slowly these days in the rapidly expanding concrete metropolis, in the middle of a large-scale development called San Francisco. The city has grown up around the neighborhood, and now acts as though it would rather not see it, like the remnants of an impoverished past, or a wound that must be healed.

Its inhabitants are locked in a drawn-out battle with the government, entrepreneurs, and politicians. They don't want to sell their land unless they get $4,000 per square meter, about $370 per square foot. A land appraisal carried out by state entities recommended less than $9 per square foot. Hungry real-estate agents and officials are lying in wait. Sometimes they show up to take photographs and measurements that no one wants them to take. Down among the shells of the old shacks, investors have glimpsed a shining pearl: undeveloped land, perhaps the last remaining in the center of Panama City.

Hens scuttle along the narrowing alleyways. Pink, aquamarine, and dark orange houses are crowded together, forming a labyrinth of balconies decorated by clothes hanging from washing lines. Addresses are scrawled on chalkboards. Five children play marbles in the cracks in the ground. You can't look up at the clouds in the sky without a tower getting in the way, looming behind the corroded zinc roofs of the shacks' first floors.

Miguel talks about fishing. "The system's changed now. For the better. Before, you'd take the product and you couldn't get rid of it. Twenty-five or thirty years ago you'd be earning ten, fifteen centavos a day."

"But aren't there fewer fish and more people fishing?"

"Yes – there's no abundance. But now the product moves and it's worth more. You sell to restaurants. You export. Sometimes the catch is poor" – he uses his longest fingernail to make a slicing gesture across his jugular vein – "so you have to go out again. No business, no life."

At night, the sea around Panama City is dotted with lamps. If you looked down from the sky, you'd see boats like Miguel's floating,

Iván Bernal Marín is an investigative journalist from Barranquilla, Colombia, now living in Bogotá. His work has appeared in the South American periodicals El Malpensante, Infobae, Semana, *and* El Heraldo, *where he also served as editor-in-chief.*

little spots of light close to the freight and cruise ships, which resemble buildings sleeping in darkness.

Batteries and light bulbs play a key role in fishing off the shores of the canal. Because if

---

If you looked down from the sky, you'd see boats like Miguel's floating, little spots of light close to the freight and cruise ships, which resemble buildings sleeping in darkness.

---

they can't see you, anybody transiting between the Atlantic and the Pacific could easily mow you down. Miguel hangs his lamps high, with bamboo canes. At night, the waves shake him unceasingly. It rains. The bursts of foam and cold winds fuse into one almighty chaos. Out at sea, his task is essentially to survive inside a washing machine going at full speed. He takes refuge in a cabin made from rods and tarpaulin. He bails out water with a bucket, his jacket and windbreaker drenched through. In the mornings he sizzles off under the sun and cooks the food he's brought along, beef and rice, on a gas stove. He looks for a cove away from the trails of the mega-ships, to avoid being caught by the Coast Guard.

Sometimes he indulges in other fare. "The sea langoustine isn't like the farmed one. Sure, you can take a chicken and tie it up and toss it some corn. But in the sea, psssh. Grab one, put it down there and see if it doesn't bite you. The other ones, the farmed ones, have no flavor. . . . They're not like the ones you get from the sea."

"So you eat what you fish?"

"Of course, man! You eat five or six first, before you sell them. That's how it is." He

marks the midpoint of his forearm. "From the sea to your plate."

Miguel leads us to Boca's quay, at the other end of the neighborhood. He has left his boat aground on the mud, tied to a stick. When the tide comes in it will float up on the spot. Four unsmiling men are repairing a net on the shore. They cut nylon, they haul, they strain, they hammer on a log. As they see me approaching, their faces seem to say: "Who is this clown?" An office chair bobs up and down in the water, with bottles, cans, bags, broom bristles, and a group of small rotting boats. This is the headquarters of one of the four traditional fishermen's associations registered in Panama City.

Day is here, coming up fast as it does in the tropics. Suddenly, our two hours' tour and series of interviews are cut off by the gestures of a bare-chested young man with a shaved head and the dead eyes of a sea bass. Tattoos reach all the way up his neck. A cross beams out from his chest. He has come from nowhere. This is not a tourist area; we unsolicited intruders know better than to mess around. We follow him towards the edge of Boca.

 Before Dawn

AT FOUR TWENTY A.M., María Vásquez returns to the shore to count her catch. It's still dark and the sea is reflecting the shoals of buildings that rise up around it. The lights from the coast caress her boat as it dances in the quay outside the seafood market. It's the last unloading of the day. A voice sings out from the radio of the man receiving her load: *"Ese toro enamorao de la luna."* The bull who fell in love with the moon.

This dark woman, fifty-five years old and grandmother to seventeen, has brought in sixteen boxes today: some sixteen hundred

pounds of colorful, meaty scallops, ripe and ready for the knife and the pan.

"Were you expecting more?"

"*Mamá*, it's just you said it was more like forty," Carlito Guzmán, her trusted distributor, replies.

"But I keep bringing them in and you sell them right off . . . that's why you can't see all of them."

Her husband Abraham, sixty-seven, is waiting for her at home. He's retired. These days their son, also called Abraham, eighteen years old and thin as a rake, is the one who actually goes out on a boat to face the waves. The woman writes everything down in a notebook while her son gets his hands dirty. When she ties up her headscarf, with its white ripples, María looks more like a pirate than a sailor. Her T-shirt is emblazoned with the shimmering English logo: "Get ready, it's time to fly." Her family is just one entry on the official register of more than eighteen hundred fishermen and women working on the shores of the canal.

Abraham Jr. starts his last job of the day by spreading the fish out like playing cards on a wooden table. This is how he counts and classifies the fruit of three days' and nights' toil on the Pacific Ocean.

The lights of a vehicle appear on the horizon every ten or fifteen minutes. An oscillating light from a lighthouse runs across the faces of the people in the marketplace. This is perhaps the only time of day transportation works. There are no traffic jams. The few cars there are cruise through the area like fish in water. Half the city is still immersed in a deep sleep, but here work began at twelve midnight, and it's only halfway done.

Megahotels, megabanks, mega shopping centers, and mega residential developments dominate the view. There are so many of them that there is no longer much space left for the fishermen. The families whose anonymous labor supplies this many-tentacled market are forced into the corners to eke out a living. They who put the wonders of the sea onto tourists' plates go about their work behind the backs of the towering edifices.

A trail of boxes, tanks, tables, and home-made scales forms a kind of reef, which spills out onto more than one lane of the road. The market is a tipping point on Balboa Avenue, the border between the Casco Viejo, the historic quarter, and modern Panama City.

If you approach the market from the shore at Cinta Costera, the highway some eighty meters away, Carlito's scales are the first ones you'll see. His stand emerges from the street corner like a peninsula. There are many more sets of scales behind it: seventy stalls spreading out towards the edges of Panama City's fishing equipment retail store. The country itself gets its name from an Indigenous word meaning "abundance of fish and butterflies." The omission of pelicans and vultures seems unjust when you notice the winged battalion keeping watch over the hustle and bustle from the rear side of the market's roof.

Buyers, fishermen, and wholesalers extend a carnival of shouts and laughter through some two blocks of wet floors, beneath a bridge covered in faded graffiti. Trucks loaded with frozen fruits of the surf line up on the corners.

Carlito wears a golden chain around his neck. Thirty-eight years old, bald and portly, he might be a Caribbean Ninja Turtle; Donatello, perhaps. The negotiations begin, the push and pull with María. The prices vary depending on the size of the overall catch that day. Today he's paying $1.25 per pound for Pacific sierra and one dollar for a pound of yellow jack, which he has quite a lot of already. In Panama, everyone counts dollars in the singular and very few use the local currency, the balboa, which they call "the Martinelli" in a candid commemoration of the president who came up with the new one-balboa coin, supposed to parallel the American dollar. They're not minting it any more.

Ahmed, Carlito's younger brother, is in charge of accounts and handles the money. He looks similar but taller, with a broader frame,

dressed in red, with the look of someone who's been awake all night because his neighbors were having a party he wasn't invited to. He would be Raphael.

They who put the wonders of the sea onto tourists' plates go about their work behind the backs of the towering edifices.

"This is a family business, my uncle started it. We're all fishermen. We have twelve vessels. We sell to whoever's buying."

 5:00 am

THE LIGHTS OF CARS and trucks are becoming more frequent, their horns closer. Emma del Mar is one of the small-scale buyers at the market. She walks among the baskets, holding bags in her hands. Her two daughters, Justine and Ayelis, drink *café con leche* by the roadside. They are laughing at the man who has just sold them the drinks. He's pushing a supermarket trolley with four thermoses in it. He only has two fangs left, slimy pincers which appear when he speaks, as if a crab were trapped inside his mouth. He charged the girls, aged ten and twelve, a quarter for each cup.

They are accompanied by their father, Agustín Menacho. He asked for his coffee "real strong, black as night." He works in construction; he builds walls inside buildings. His wife Emma fries fish and sells them at her house, in the 24 de Diciembre neighborhood, for six dollars a plate. "A platter of mixed seafood, fish, and plantain. That's how we get by," she says. She's bought seven dollars' worth of corvina, he a fillet of seabass for lunch.

The couple embodies two contrasting activities, which both contribute to Panama's GDP. Construction is one of the sectors that have seen the largest growth. It's close to making up 15 percent of the national economy and brings in more than $11 billion per year. Fishing, on the other hand, is in freefall. It has gone from 3 percent of GDP in 2003 to 0.3 percent in 2019, when it brought in $187 million, much less than the $350 million it was worth just four years previously.

None of this bothers Emma and Agustín. They are only concerned about the fifty minutes, or more, that it will take them to travel home in their car. Especially if they continue to be unreasonably delayed by curious journalists.

 **5:20 am**

"I'M HERE TO CLEAN for all you poor and humble souls!" shouts an Indigenous man, wearing a vest and a knitted hat which reads "Jordan 23." The vehicles are now forming a river of horns and roaring motors.

He holds out a fish in one hand and scales it with the other, like a *salsero* scratching a rhythm on a *guacharaca*. He piles up the cleaned corvinas on the table, in the middle of the road's right lane. Now the cars can only use the left one. A breeze shakes everything as they accelerate past. "No parking here," the wall behind him says.

"Our work is surrounded by danger. A crazy guy might come along and then *boom*! They say terrorists are crazy, but not as crazy as the people who work here." He identifies himself as Alcibíades, "The Scraper." He comes from the island of Ailigandí, in his native community of Guna Yala. He puts the eleven fish he has just scaled for $1.60 into a bag. He prepares them so that they are ready "to put

on your plate and eat right now." He removes the belly; he uses the scraps to make remedies for cancer.

There's a local saying: "*más salsa que pescao*," "more sauce than fish," roughly equivalent in meaning to "all talk, no action." But Panama is more fish than salsa. It's covered by a thin, piquant layer, the one you see on postcards; it's the enchanting land that gave birth to the *salsero* Rubén Blades. But there's far more flesh beneath the surface. (And, with the exception of the dock area, in most places these days you'll hear far more reggaeton than salsa, anyway.)

In recent years, an annual average of two hundred fifty thousand tons of fish has been docked in Panama. That figure combines the anchovies, herring, pacific bumper, tuna, shark, langoustines, giant hawkfish, and shrimp from industrial fishing with the porgy, Pacific sierra, seabass, yellow jack, prawns, octopuses, and crabs caught by traditional fishermen. That's what it says in the censuses carried out by the audit office, but far more goes unreported.

"In kitchens, in restaurants, it's expensive, sure. Here, it's cheap. It's cheap for the fisherman. It's the reselling that does that." María Vásquez knows that her sierras will get pricier and pricier the further they get from the sea floor. They're longer and thicker than baseball bats.

 **6:00 am**

THE CARS OUTSIDE are slowing down. The night and its accompanying hustle and bustle slip away through the hands of daytime and its accompanying traffic jams. Panama City boasts not only two oceans, but two suns; one emerges from the Bay of Panama and the other is reflected in the façades of its skyscrapers. A

van bearing the insignia of the Institutional Protection Service (SPI) cries out on its megaphone: "Please move to where the safety of others will not be compromised!" The SPI agents have decreed that the market's working day ends at six thirty, when morning comes and the moon slips into the sea. The officials move the fishermen off the road and shift the scales and tables over into each stall. The pelicans descend from the roofs to feast on fish scraps on the ground. Bursts of red and gold emerge from the clouds.

The first of the market's restaurant shutters rise, on the side that faces the sea. The morning breeze greets queues of tourists in phosphorescent shorts and floppy hats. The palm trees sway coquettishly in time with the acoustic arpeggios of yet another salsa classic, beloved of grandparents and fishermen: "Every single night, my sweetheart, I spend it sleepless, my love . . . thinking of you . . . aching for you."

The radio is still playing in Carlito's stall. On the sea-facing side, where the fish is served cooked, he uses the name Econofish. For $15 Econofish serves up a crunchy fried piece of porgy. The price does not include the land dispute the fishermen are waging against the big construction companies, or the tension with the authorities over occupying public space in the early mornings. What it does include is plantains and a green salad. No one cares about any of the rest.

 **Get Out**

WE HURRY BACK through the alleys. The tattooed young man from the fishermen's

association escorts us through the passage-ways, scowling as he walks.

"Come on now, guys, you're not stupid," he says. "If you're going to talk, talk up there. No taking photos here. Off you go." He says it in a tone that no one challenges. As he walks, his mouth and eyes, trembling but never shut, recall those of a catfish, pulled out of the water and floundering on land. He looks back, then to either side, then back again, as we scamper to the edge of the neighborhood.

In Boca la Caja you can see two cities in one. Many prefer the farmed Panama over the wild, maritime variety. The one that is incubated in post-capitalism and brimming over with invest-ments from tax-evading foreigners and shell companies. That smells of First World perfume, not Third World fish. That doesn't bite.

This neighborhood has a reputation for smuggling and drug trafficking. Telling the locals that you're a journalist won't get you very far. But Miguel the fisherman sees no problem in continuing to talk to us from his place back near the handwritten sign at the very edge of Boca. He sits on a sidewalk, using another bottle as an anchor. That is, until a new voice repeats the warning, louder this time.

"Hey, get those guys outta here!"

The shout comes from a young man sporting a black cap and a curly Van Dyke beard. He stops behind an incongruous brand-new car with darkly tinted windows, keeping his hands out of sight. Only his chest, arms and shoulders, sinewed and tensed, are visible. His mouth stays open. He is looking straight over, without blinking. It's as if a deadly serious version of Mario Balotelli's angry goal celebration were staring right at you. Someone else points at us with a harpoon finger. Miguel falls silent, and with a hand gesture he makes it clear that we should leave. We know the neigh-borhood's reputation. It's not worth it. Time to get out of here.

The Spanish original of this article first appeared in El Malpensante *number 194, March 2018. Translated by Rahul Bery. Official data has been slightly updated in the English version.*

Len Stomski,
Greylock from 116,
oil on gessoed panel

The Berkshires

The Appalachian ranges are believed to be the oldest
mountains in the world.

Once jagged peaks, they're now but rolling hills,
more welcoming than when they were sublime.
The trickling water pleases where it spills,
where craggy peaks give place to rolling hills,
easing the mind like Wordsworth's daffodils.
An older mountain's easier to climb.
What once were jagged peaks are rolling hills,
more welcoming this way, if less sublime.

ALFRED NICOL

The Abyss
of Beauty

IAN MARCUS CORBIN

Why do we shy
from paying
attention to
nature's beauty?

*A*lbert Camus writes that if you're truly paying attention, beauty, for all its sweetness, is "unbearable." Beauty, he says, "drives us to despair, offering us for a minute the glimpse of

Previous spread: Peer Christensen, *Crabapple Study,* oil on canvas, 2020

an eternity that we should like to stretch out over the whole of time."

For most of us, Camus's pronouncement sounds dubious; it has the ring of tragic poetic fancy. We may feel revulsion or despair at the sight of misery and death, but beauty? What sort of pain could attend the apprehension of a sunset or a flower?

And yet it's easy to be unsure if he's correct, because true looking is rare. Our customary mode is to look *for* things rather than *at* them, to register them just long enough to tell whether they'll harm or help, what we'd better steer around, what we should pick up from the ground and pocket for tomorrow.

One afternoon last summer, I was sitting on a bench in a small urban park, my youngest son Leonard asleep in his stroller. I'd consciously chosen to leave my iPhone at home, determined to look around me as I went. It's an ongoing ethical project, a way of life I aspire to and too rarely achieve. I have a running suspicion that I could really, deeply love life, or a day or afternoon at the very least, if I could just be quiet and look, stop the incessant scheming and worrying and mental grappling. When Gerard Manley Hopkins sits still, he finds that the natural world is "charged with the grandeur of God," and exults in the knowledge that its "blue-bleak embers" "fall, gall themselves, and gash gold-vermillion." That's what I want. I want to see embers, blue-bleak and dying, to *see* that when they fall and gall themselves, gold-vermillion gashes out into the visible world. How different would that be from my current life of cars and sidewalks and text exchanges, of

Ian Marcus Corbin is a writer and philosopher in Cambridge, Massachusetts. He is currently co-director of the Human Network Initiative at Harvard Medical School and a Senior Fellow at the think tank Capita.

Artwork by Peer Christensen. Used by permission.

Bird photographer Rufus Wareham spent much of his youth roaming New York's Catskill Mountains. He currently lives at the Danthonia Bruderhof in New South Wales, Australia, sighting approximately four hundred bird species each year.

Rufus Wareham

Mulga Parrot – Queensland

"The mulga parrot is one of many elusive species found deep in the Australian outback. On a business trip to Queensland, I visited a wildlife refuge in hopes of seeing some of these singular birds. I couldn't believe my luck when one landed nearby. This is a female; her colors are more subtle than the flamboyant male's."

long nights in my restless, thought-infested bed? Perhaps we can *see* ourselves to life.

On the other hand, maybe what we'll see if we truly look is cruel, unbearable. In the little park my eyes came to rest on a tree whose branches stretched in over the chain-link fence from an uncultivated little area outside. It was August and the leaves were thick and vibrant, lusty almost, and a hot afternoon sun lit them from behind. On the ends of its branches hung thick clusters of pea-like orbs, some sort of reproductive vehicle, bearers of the next generation of stubborn urban weed-tree. The normalest thing in the world.

But deprived of anything more urgent to look at, my eyes grew settled and quiet, the tree's announcement of itself grew louder. It became marvelous. That heavy, fruitful, virtuosic hanging; tucked in this ignored corner, unattended, these rich, bursting clumps of waxy green generation. So much plenitude there, so much gratuitous newness, a production of

far more fullness – *new life* – than I could ever fashion with my busy mind and hands.

I have, as should be clear, Romantic tendencies, and I was enthralled. So much, so much on this little unplanned burst of branches. I had no practical interest in the seeds, or whatever they were. But I wanted to keep them. It was too much beauty to let go with indifference. That rich plumpness – what a thing to exist. I marveled at this unaccountable, unnoticed beauty. I wanted to have it forever, to live somehow in its presence. Life, I felt strongly, is better than I normally live it.

And yet, what the hell could it mean? What could I do? These little seeds were here for a week or a month; they would probably not succeed in making another tree. They would fall and shrivel; by now they are certainly rotted into dirt. I felt, suddenly, a pang of dismay, or even despair. I was too far from them; I can hardly recall now what they looked like. Seeing them felt like a torture, a tease. As

I walked home, I was reminded of "Fruit," by one of our great modern guides to these things, the Polish poet Adam Zagajewski:

> *. . . How unattainable*
> *afternoons, ripe, tumultuous, leaves*
> *bursting with sap; swollen fruit, the rustling*
> *silks of women who pass on the other*
> *side of the street, and the shouts of boys*
> *leaving school. Unattainable. The simplest*
> *apple inscrutable, round.*
> *The crowns of trees shake in warm*
> *currents of air. Unattainably distant*
>   *mountains.*
> *Intangible rainbows. Huge cliffs of clouds*
> *flowing slowly through the sky. The*
>   *sumptuous,*
> *unattainable afternoon. My life,*
> *swirling, unattainable, free.*

What a strange kind of animal we must be, to feel ourselves perched on the periphery of something, always only almost living. The

What a strange kind of animal we must be, to feel ourselves perched on the periphery of something, always only almost living.

thing in front of us, just a hair past our reach, seems ideal, if we could get to it. But can we? Does such intimacy exist? Are we delusional to hope so?

Camus knows, and is sad but brave. Hopkins knows different, and is full of gratitude. For him every small cluster of berries is charged with spiritual grandeur, because "the Holy Ghost over the bent / World broods with warm breast and with ah! bright wings." Camus, too, seems to have glimpsed some shadow of divinity around the skirts of material beauty – it's precisely an evasive, illusory "eternity" that hurts him, because he knows he cannot make it stay. Zagajewski also sees this shadow, but he doesn't really know, not like Camus and Hopkins do. His is a poetics, an aesthetics, even a spirituality that is charged with longing and generosity – knowing little, refusing to give up hope, relentlessly honest. I find myself, in fits and starts, there with him.

BUT NOT EVERYONE IS stuck in limbo. Some people know, or at least believe. But how? I have spent well over a decade poking at these questions with the tools of philosophy, and I know at least that arguments can't be the answer. Perhaps we can begin to sight a map by looking at the story of Václav Havel, the Czech writer and intellectual who was influential in the world of underground arts and letters in Soviet-Bloc Czechoslovakia. In 1989, he helped to foment the bloodless anti-Soviet "Velvet Revolution," and subsequently became president. Ten years before, Havel had been arrested on charges of subversion, and sentenced to a four-year prison term. One day in the courtyard of the prison at Heřmanice, he had a dramatic epiphany that amounted to something like a conversion experience. He wrote to his wife, Olga:

> *I call to mind that distant moment in Heřmanice when on a hot, cloudless summer day, I sat on a pile of rusty iron and gazed into the crown of an enormous tree that stretched, with dignified repose, up and over all the fences, wires, bars, and watchtowers that separated me from it. As I watched the imperceptible trembling of its leaves against an endless sky, I was overcome by*

Rufus Wareham
Common Loon – Adirondacks

"I was paddling my canoe across an Adirondack lake when I was joined by a family of common loons. The mother dove for small fish and other delectables, dropping them down the open beak of the waiting baby. They kept even with my canoe, sometimes even swimming right under the boat. When they glided into the shadows under some trees, one gave me a chance for a picture that highlights the beauty of its feathers."

a sensation that is difficult to describe: all at once, I seemed to rise above all the coordinates of my momentary existence in the world into a kind of state outside time in which all the beautiful things I have ever seen and experienced existed in a total "co-present"; I felt a sense of reconciliation, indeed of an almost gentle assent to the inevitable course of events as revealed to me now, and this combined with a carefree determination to face what had to be faced. A profound amazement at the sovereignty of Being became a dizzy sensation of tumbling endlessly into the abyss of its mystery; an unbounded joy at being alive, at having been given the chance to live through all I have lived through, and at the fact that everything has a deep and obvious meaning – this joy formed a strange alliance in me with a vague horror at the inapprehensibility and unattainability of everything I was so close to in that moment, standing at the very "edge of the infinite"; I was flooded with a sense of ultimate happiness and harmony with the world and with myself, with that moment, with all the moments I could call up, and with everything invisible that lies behind it and has meaning. I would even say that I was somehow "struck by love," though I don't know precisely for whom or what.

This epiphany grants to Havel what he describes as "a sense of ultimate happiness and harmony with the world and with myself, with that moment, with all the moments I could call up, and with everything invisible that lies behind it and has meaning." But how, why has this happened? What is it about this particular tree – and by extension, any ordinary beautiful thing – that might explain this sort of experience?

We can't finally know, of course, but we would do well to begin with Havel's description of the tree. The tree is "enormous," stretching out its branches with "dignified repose, up and

Peer Christensen,
Tamarack Study,
oil on canvas,
2018

over all the fences, wires, bars, and watch-towers" as its leaves "tremble imperceptibly" against "an endless sky." His epiphany seems to be fostered by the juxtaposition of the tree and sky, each illuminating his apprehension of the other.

He sees living things stretch out with "dignified repose," in their ceaseless, simple, unreflective excellence. Zebra fur just *does* grow in stripes, cuts heal, magnolia trees bloom in the spring. Within the simplest cell, structures and functions of great complexity and elegance operate independent of human comprehension or control. And this sort of virtuosity is beautiful to behold, as Aristotle observes: "Absence of haphazard and conduciveness of everything to an end are to be found in Nature's works in the highest degree, and the resultant end of her generations and combinations is a form of the beautiful."

This beautiful virtuosity goes further than we can imagine, perhaps infinitely further. Havel is struck by nature's "own great and mysterious order, its own direction." Our sciences, especially physics, aim to map out an explanatory basement, an account of the forces that finally underlie the structure and order of trees, hands, and eyelashes. But even if we embrace string theory we can always wonder what accounts for the cohesion and function of their tiny vibrating strings of energy. Living things have depths which are simultaneously orderly and incomprehensible to human knowers.

Confronted by mystery, Havel is dizzy at his own finitude. He sees – with relief – that he simply cannot make himself omnipotent, nor should he try. And so he describes a feeling of "gentle assent to the inevitable course of events as revealed to me now . . . combined with a carefree determination to face what had to be faced."

Rufus Wareham

Golden Whistler – Elsmore, New South Wales

"You can hear the golden whistler's song in our east Australian woodlands. Last year, I was passing a house just as a whistler flew into a window, stunning itself momentarily. I lifted it gently off the ground, and, as it began to stir, placed it on a nearby branch, waiting till its small claws took hold. I stepped back and snapped the photo. Within a few minutes, it dashed off."

Still he feels "a vague horror at the inapprehensibility and unattainability of everything I was so close to in that moment, standing at the very 'edge of the infinite.'" Havel has come to see that the human quest for total understanding and mastery is destined to remain frustrated. As parts of nature, we humans contain the same infinite depths as Havel's enormous tree, but as knowers and actors, we are incorrigibly finite. This bittersweet realization makes Havel receptive to the marvelous insight that if we can't understand everything, maybe there's *a lot* we can't understand.

Each tree, each ant and person rises and falls beneath an infinite sky; we may flourish on this day or that, but the passage of time will see to our physical annihilation. The virtuosity of life is all the more striking – it might even seem *miraculous* – when it emerges in a context which also foregrounds its fragility. Here we all are, for a few short minutes – tiny, brittle, ignorant, and unspeakably beautiful. Confronted

with this juxtaposition, Havel suddenly, unexpectedly receives the intuition that behind the façade of ubiquitous flux, a limitless, incorruptible reality exists. This is the reality that traditional metaphysics and theology try to talk about.

Havel gratefully accepts his intuition, and his posture towards life is transformed, though he does not go so far as to label himself a convert: "I would even say that I was somehow 'struck by love' though I don't know precisely for whom or what." Like a number of theologians, Augustine and Aquinas among them, Havel senses the difficulty of attaching words to our experience of the ineffable.

And yet, unlike them, he feels unable to accept any doctrinal attempts to do so, though he does say in other writings that he wishes he could. He writes in another letter to Olga that "by perceiving ourselves as part of the river, we accept our responsibility for the river as a whole." Havel's aesthetic rapture doesn't whisk him away from the bonds of human solidarity, but reshapes and strengthens them.

Faced in that moment with the juxtaposition of fragility and virtuosity, Havel suddenly knew that fragility was not the final word. Particular people and things are indeed fragile,

❧

We are left with one main imperative: humble, careful attention.

but there's some ineffable spiritual reality that unites and harmonizes the particulars into a beautiful, invulnerable whole. The universe is not, he saw, cold and indifferent.

Why on earth did Havel come to this conclusion? Though "conclusion" is the wrong word: this was not the result of long consideration or weighing of evidence. Havel stepped like an ordinary prisoner into a dusty courtyard, and suddenly metaphysics bubbled up, unbidden. And he was swept away by its truth.

T O UNDERLINE AGAIN: Havel was not logically forced to his conclusion. We can marvel at nature's virtuosity even if fragility is indeed the last word. But to Havel, again, it did not *seem* to be the last word. The perception of a moment of beauty snowballed into an intuition about beauty itself, that gathered and pulled all of existence into its orbit. But if beauty sometimes seems to open a small window to a realm beyond our fragmented, time-bound existence, every adult knows that things are not always how they seem. Herman Melville, writing in 1851 to his friend Nathaniel Hawthorne, says:

In reading some of Goethe's sayings, so worshipped by his votaries, I came across this, "Live in the all." That is to say, your separate identity is but a wretched one – good; but get out of yourself, spread and expand yourself, and bring to yourself the tinglings of life that are felt in the flowers and the woods, that are felt in the planets Saturn and Venus, and the Fixed Stars. What nonsense!

But he adds,

N.B. This "all" feeling, though, there is some truth in. You must often have felt it, lying on the grass on a warm summer's day. Your legs seem to send out shoots into the earth. Your hair feels like leaves upon your head. This is the all *feeling. But what plays the mischief with the truth is that men will insist upon the universal application of a temporary feeling or opinion.*

For Havel, the *all* feeling was a glimpse of a genuine truth, and it transformed him. For Melville it was a temporary "feeling or opinion," to be enjoyed but not to be trusted as revelatory. Why did Melville "know" that the seeming reality of the "all" was to be rejected as fancy? Why did Havel "know" that it was to be embraced?

As thinkers – appliers of concepts, technicians of logical implication – I fear that our line of inquiry has to end here, in disappointment, with these questions. All we can say for sure is that some experiences of beauty make the world seem a certain way – as if it is "charged with the grandeur of God," as Gerard Manley Hopkins puts it. On the other hand, to his contemporary Matthew Arnold, the world seems to be a "darkling plain" where "ignorant armies clash by night."

Deciding which understanding to embrace as true cuts so close to our fundamental

Rufus Wareham

Osprey – Esopus Meadows

"It's breathtaking to watch
an osprey fold its wings and
plummet into the water
at speeds of up to eighty
miles an hour, then struggle
back into the air with a fish
writhing in its talons. Esopus
Meadows, a small park along
the banks of the Hudson
River, is a favorite hunting
spot for these dive-bombers.
In the light of an early August
morning, this one took off
directly overhead."

experience of reality that it's hardly a decision at all. Philosophers dig, Wittgenstein says, until we hit rock, and the spade turns. And the spade turns here. How, on what grounds, could Havel convince Melville to believe the *all* feeling? How could Melville convince Havel to disbelieve it? We are not in the realm of logical confirmation or refutation here, or even of persuasion. The question is closer to one of trust. Havel and Melville are perhaps comparable to a young groom receiving a pledge of love and fidelity from his would-be bride. Should he believe her? Only he can make that decision – there's no *a priori* way to adjudicate it. Some experiences can only be lived through, and listened to.

So we are left with one main imperative: humble, careful attention. This is the only way to begin to examine the great question Havel and Melville answer in their own, radically different, ways. The sort of attention that can approach the question is perhaps what Simone Weil describes as "empty, waiting, not seeking anything, but ready to receive in its naked truth the object that is to penetrate it."

Weil's image comports very uncomfortably with the aggressive, penetrating style of attention that took pride of place in the Enlightenment. This style still reigns as the paradigmatic posture of serious inquiry, vindicated every day by the fantastic advances of modern science and technology. It is perfectly articulated in Kant's *Critique of Pure Reason,* when he says that reason "must approach nature in order to be taught by it. It must not, however, do so in the character of a pupil who listens to everything that the teacher chooses to say, but of an appointed judge who compels the witnesses to answer questions which he has himself formulated." This is, of course, a rather cramped and narrow way to approach nature, at least when we act in our capacities as mere humans rather than scientific investigators.

The Kantian approach has steadily become

the default way of experiencing the world, in the process helping to make experiences like Havel's rarer. Kant's approach has its time and place, to be sure, but some topics – like the fundamental nature of reality, or the likelihood

Openness requires a degree of brokenness, and may be difficult and painful.

of marital concord – require a more open, variegated, sensitive approach.

This openness, though, requires a degree of brokenness, and may be difficult and painful. It is not without resonance that Havel's epiphany took place while he was imprisoned and powerless – Havel's awareness of his finitude, his confrontation with a world that is too great for his capacity, is reinforced by physical confinement. It is difficult to adopt Kant's posture when one's illusion of mastery has been so unceremoniously and convincingly destroyed. It makes sense, then, that thinkers from Boethius and Thomas More to Wittgenstein, Pound, Dostoyevsky, and Martin Luther King Jr. have penned masterpieces from prison. The power of suffering and humiliation to open our eyes is a guiding theme of some of Dostoyevsky's most powerful work, which is to say, some of the most powerful ever created.

WHEN THE WRITER and atheist polemicist Christopher Hitchens was first diagnosed with cancer, he announced that if he should, in the grips of decline, begin to soften his stance toward theism, the public should ignore him.

This Christopher Hitchens, he said – still robust, pugilistic, lustful – was the real one. The diminished, sad, terrified Christopher Hitchens of advanced-stage cancer would be just a shadow of the full man. Hitchens here has broached a massive epistemological question – do we see more clearly in our weakness or our strength? Dostoyevsky answers definitively for the former, but with a qualification: the kind of weakness he finds illuminating involves a careful distinction between guilt and shame.

We can see this distinction summed up in the story of Markel, a minor character in *The Brothers Karamazov*. Markel is a haughty, cynical boy, who mocks his mother's piety, until he is struck with illness. Then he is progressively weakened, progressively humbled; he comes more and more to embrace the piety he once mocked. Speaking to his mother, Markel calls her "heart of my heart, my joyful one," and tells her to "know that verily each of us is guilty before everyone, for everyone and everything. I do not know how to explain it to you, but I feel it so strongly that it pains me." From his new perspective, Markel turns in humble repentance toward the people who surround him and toward the broader creation: "There was so much of God's glory around me: birds, trees, meadows, sky, and I alone lived in shame, I alone dishonored everything, and did not notice the beauty and glory of it all."

Shame and guilt here are polar opposites. Shame is denial, a desire to hide one's true nature from oneself and others. It is what Hitchens imagined religion might be – a scrambling, pathetic attempt to escape finitude. The guilt that Dostoyevsky has in mind is the opposite. It is a letting go, the generous acknowledgement of what one truly is, the hospitable acceptance of one's own flaws, finitude, mortality. This acceptance liberates

Rufus Wareham
Red-headed Woodpecker – Plutarch Swamp

"A few swamps in the mid-Hudson Valley are known locally for their population of red-headed woodpeckers. I climbed a small tree on a swamp edge and opened a view through some branches. Then – I waited. Five or six of the birds flew around, calling in high-pitched screeches as they collected bugs and moths. But no luck. Just as I was about to leave, one landed only twenty-five feet away. It perched for a total of two seconds. Click! Sometimes three hours of patience are rewarded."

a person from stifling falsehood and instead creates the possibility of reunion: with oneself, with others, with nature. Markel's mother tries to save him from this liberation, assuring him that he is not so guilty as he claims, and he reproves her: "Let me be sinful before everyone, but so that everyone will forgive me, and that is paradise. Am I not in paradise now?" These are heavy, fragrant words.

From some deep part of me I feel their affirmation; if there is a key to my overheated encounter with some pea-like tree fruits in a local park last summer, it would seem to reside here. Reading through the lenses of Havel and Dostoyevsky, it seems we could say that my hands and eyes were too weak to grasp what I saw – I could neither hold the fruits in perpetual existence, nor see to their core, to understand and love the force that animates them. Dostoyevsky suggests that my lack of control is the simple human reality, and my anguish at it a blinding, crippling form of shame, perhaps the great human disease. He suggests that I could have kept looking, I could have kept feeling my powerlessness. This sort of fortitude, in the face of my limitations, might have led me to a posture of guilt; it might have led me to understand what I am and to embrace it as good. From there, some large things – far too large to put into words – might have presented themselves to me.

Dostoyevsky asks a lot; it sounds like hard, hard, harrowing work. It sounds like being shaken to the core in a small grubby park, with nannies and yuppie parents all around checking their phones and herding their little strivers. It sounds, maybe, like more than I could take. It also sounds, however, like the way reality is glimpsed – whether that reality be crushing or saving, eternity or death.

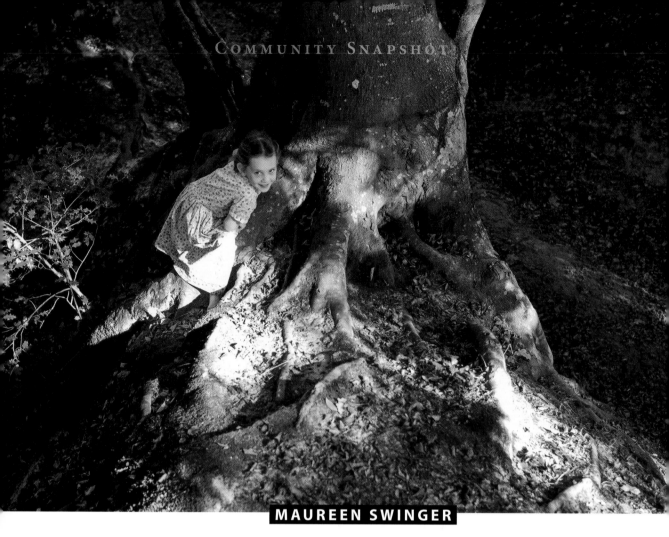

MAUREEN SWINGER

My Forest Education

We were a group of unteachable second graders –
until our teacher introduced us to the woods.

MY FIRST YEAR OF SCHOOL smelled of cigarettes. The classroom was a box – American flag front right, window front left. (I still think of rights and lefts in flags and windows.) Most of my time was spent looking left; there was a swallow's nest in the ivy on the next building. The teacher was a little annoyed with left-handed kids. Those are almost the only memories I can pull to the surface when my first grader asks me what first grade was like. But second grade now . . . that was spent almost entirely in Piney Woods. "Where's that? Is it a storybook place?" Well, yes, child. Here's the story.

Maureen Swinger is an editor at Plough *and lives at the Fox Hill Bruderhof in Walden, New York, with her husband, Jason, and their three children.*

In the Allegheny Mountains of Pennsylvania, on the edge of a hilltop clearing at New Meadow Run Bruderhof, someone long ago planted acres of white pines in perfectly straight rows. I have no idea of their intended purpose, but to a kid whose family had just moved from a crowded house on a steep city street, their fated purpose was freedom. Darting between sun-dappled trunks whose lowest limbs started just above your head; running barefoot, soundlessly, on untold layers of soft pine needles; flitting from shadow to shadow in stealth raids between the camps of pirates and princesses (fierce princesses); building ever-expanding, interconnected tree forts; collecting fallen branches for camp-craft fires; roasting doughboys, baked apples, s'mores . . . what memories don't I have?

We had a classroom too. I can picture it if I try: walls full of maps, bird lists, weather charts. We put in a full and enriching academic year. It's just that the outdoor learning was richer.

Bruderhof schools emphasize science, nature, hands-on exploration. It's rare to find an afternoon class indoors in any weather. But our class took it up a level, thanks to sheer contrariness. Thanks, in fact, to a teacher who knew what to do with contrariness.

Our crew – well, let's just say we had left several student teachers questioning their calling. With just sixteen kids it should have been a manageable class, but a bunch of us were still finding our way into community life, or navigating learning or behavioral difficulties, or just edgy and argumentative by nature.

DICK WAREHAM WAS SIXTY-ONE years old; his health wasn't great; he had been taking a break after decades of teaching. I don't know if he offered to take us on or was begged to. But the day we found ourselves at the top of the clearing, facing a big man wearing a

red bandana and palming a stopwatch, we all shut up and fell in line. Whatever was going on, it looked more exciting than fighting.

Dick had created an obstacle course that made use of every manmade and natural object in a half-mile loop of field and woodland, with a pavilion and playground thrown in for good measure. Stations were numbered with small wooden tags. Instead of a direct competition, we were to run against ourselves, building our speed and agility over the course of the year. Good news for one uncoordinated asthmatic.

On your mark – get set – GO! Shinny up and down the steel swing-set frame, leap in and out six open windows of the pavilion, scramble up the stone chimney and slap the number 3 tag at the top (jump from halfway down to save two seconds), sprint for the seesaw, drop and roll under the frame (dang it, snagged your shirt), teeter along the balance beam (double back if you lose your footing), take a run-off and vault the fence into Piney Woods, weave between alternating pine trunks at top speed for the length of the line (see how straight you can run when you come out), then it's all engines firing for the finish line, as you swerve past the site of our sapling lean-to clubhouse. Every time you run this course, you'll feel a small combustion of pride in the rising walls and roof beams, just as your oxygen levels are flagging.

Behind you, before you, at one-minute intervals, friends and foes are flying along the same track, pushing for the extra burst of speed that will throw them over the finish line in a wheezing heap of achievement.

And it *was* an achievement, as we found out at the end of the season when we invited our dads to run the course, with mixed and entertaining results. Dick charted all our times, and it was no secret that one particular blur of speed was always going to clock in around 5:07, but the rest of us were delighted to hear our ever-improving times announced with Olympic gravitas as the stopwatch punched down. Any incremental advance was celebrated. Nobody had to worry about occupying the last time slot, because that line was already filled. The final listing in the season's chart read: "Dick W: 28 minutes, 365 seconds. Thanks everybody! They were all very good times, except the last one. He needs to shape up a bit!"

NATURALLY WE HAD NO IDEA that this big, slow grandpa who needed to "shape up a bit" had been voted an all-star basketball player; his college team had traveled the country facing off against schools far beyond their class. The scouts were watching, and Dick was offered a contract to play professional basketball. In the 1940s, this was the forerunner of the NBA leagues.

He could have followed his basketball dream. Instead, he walked away from the offer and enrolled in Bethany Theological Seminary in Chicago, convinced that God wanted more of him than a good game. But the seminary couldn't see any point in burying talent, and made him its athletic director and coach of the basketball team. At the same time, he worked as a Church of the Brethren youth counselor. The teens who were having too much fun disrupting the church services were met with humor, a calm faith, and a hoops challenge.

We know about this history, because many of those former church service disrupters joined the Bruderhof movement later in the 1950s, along with Dick and his wife, Cosette, whom he had met at the youth group gatherings. In fact, one of Bethany's best young basketball players was Glenn Swinger, later (quite a bit later) my husband's grandfather. When they all got back together to hang out, their memories bounced back and forth like a high-speed give-and-go.

But eight-year-olds don't bother much with what the old folks did "back when." All we knew was that life was mighty fine. Running separate races together had shaped us into a team: less likely to tangle with each other for the fleeting satisfaction of starting a fracas, more likely to fall into close formation around the teacher as we loped up the steep hill to "our" clearing. We loved this man who never raised his voice but always kept us listening.

Staying together wasn't mandatory, but if you were dawdling in the back, you might not discover why cirrostratus clouds could create a halo effect around the sun, and what sort of weather they foretold. Chances were you'd miss the scarlet tanager's dash between the pin oaks. You wouldn't know you could eat these wintergreen berries, or use this shredded birch bark to start a fire even if the wood's a little damp. (It's the oil.) If a meteor shower was on the sky calendar for that night, we'd be the only ones hassling our parents to come out and watch with us. "Perseid what?"

Map and compass reading, tracking, weather watching, outdoor cooking, getting along with fellow humans – it all seemed to simply happen as we went about our days. But I can remember it now effortlessly, while much that was studied in the intervening years has long been absent without leave, or care.

COMMUNITY LEGEND HAS IT that our clearing was called "The Clearing" because the man who had a vision for a small picnic space among the trees hired a local bulldozer operator to clear some ground until he was told to stop. Unfortunately the visionary was called away on a sudden trip, while the bulldozer operator continued his stoic deforestation for quite a few too many days. This created our beloved wide clearing, with ample play and racing space, plus the necessity for some "clearing up" of the situation upon the return of the traveler.

The clearing project happened long years before we were to benefit, but that wonderful bulldozing mountain man was alive and well when our class went on a day trip to his homestead. He showed us his pit full of timber rattlers. "They're ma' pets," he assured us. "They're friendly as can be."

"Yep, friends," said Dick, as we trooped back to the van. "Till somebody looks sideways at someone and then it's all rattles and fangs." I still can't think what he was hinting at.

If home base was the Clearing and Piney Woods, we roamed all over from there, visiting neighbors, studying Revolutionary War battlefields and fort ruins, searching for fossils at old quarries, swimming at the Youghiogheny Dam. On "free play" afternoons, it was up to you what you did with your stretch of wilderness. Build dams, catch crawdads, whittle something unrecognizable – Dick didn't care, as long as you told him where you were going, and he could find you there.

He always knew where my friend Liz and I were. Near the stream that ran past the meadow below the school, an old apple tree on a bank had a split trunk and big level branches, one for each of us. We sprawled there like literary leopards, a stack of books balanced within arms' reach. A wind-tousled tree is the best place to revel in tales of adventure on the high seas. (Green apples make an excellent defense against approaching marauders. Also, deck provisions.)

THAT APPLE TREE is still standing, though the school building isn't. I don't travel back to my childhood home as often as I'd like, and when I do, I try not to exclaim, "Oh, this is gone! That is changed!" I hate hearing such laments from others; why should time freeze just because we were happy somewhere? All the same, I'm so relieved that the old tree hasn't gotten cleared; it looks *exactly* the same, and I half expect a small green apple to come whizzing down through the leaves – courtesy of a shade of childhood perhaps, piqued at being interrupted mid-chapter. But then, my aim was never that good.

Why Children Need Nature

Friedrich Froebel

If we are fully to attain our destiny, so far as earthly development will permit this, if we are to become truly unbroken living units, we must feel and know ourselves to be one, not only with God and humanity, but also with nature.

Consequently, parents and family should regard contact with nature as one of the chief moving forces of the life of the child, and should make it as full and rich as possible. And the best means is play, for at first play is the child's natural life.

The human being, especially in childhood, should become closely acquainted with nature – not merely with its details and forms, but with the divine spirit that is contained within it. This the child needs and feels deeply. Where this sense of nature is still unspoiled, nothing unites teacher and pupil so closely. . . . Teachers should regularly take their classes out of doors – not driving them out like a flock of sheep or leading them like a company of soldiers, but walking with them as a father with his sons or a brother with his brothers and making them more familiar with whatever the season offers.

We see many adults who have grown up amid scenes of natural beauty and yet are unconscious of their charm. Children feel themselves drawn towards the spiritual in nature, but unless this yearning is welcomed and strengthened by their elders, either it dies away or they lose their confidence in those whom they should respect. That is why children and adults should go out together, and together strive to feel in their hearts the spirit and life of nature. ➤

Friedrich Froebel, who created the concept of the kindergarten, was a reformist nineteenth-century educator who emphasized the value of teaching the "whole child" through active play, creativity, music, art, and hands-on learning.

Piney Woods, though, is an impenetrable wall of tangled undergrowth and dead limbs. There's no use getting maudlin. White pines don't live forever, and the children of this community will never run out of places to play. Why should I feel as if behind the wall, fierce princesses and pirates are sleeping for one hundred years, when we're all grown up and have small pirates of our own to contend with?

DICK WAREHAM DIED IN 2001, age seventy-six, from cancer. In his last days, many of us had the chance to stop by, spend time with him and Cosette, and thank him for being the teacher that he was. But the best tribute I saw was an impromptu one. Later that year, New Meadow Run hosted a youth conference. More than a hundred young people from a dozen communities took an afternoon hike across the Clearing and along the edge of the woods. Liz and I found each other and drifted toward the back – not the last, but almost. I can't remember which one of us said, "If the number three tag is still there, let's hit it."

As we looked ahead to the old pavilion, someone broke out of the herd, scrambled up the stone chimney, slapped the top, and jumped down. A few moments later, someone else followed suit. We saw grins, shrugs, questions from their companions. I don't know if they were given explanations. By the time we got there to find the familiar worn hand-and-foot-holds (but only need half of them), seven others had made the vertical pilgrimage. As far as I know, we were all the alumni at the conference.

In the end, I can't grieve for an old forest – it's right there when my children ask about my childhood. The afternoon sun is slanting through the long, green aisles, the needles are soft underfoot, the air smells of pine pitch, and we're laughing as we run back toward the big man with the stopwatch who is always proud of us, whatever time we make. ➤

In the Image of God

J. Heinrich Arnold

WHAT IS MAN? He exists in a field of tension between animal and spirit. Most people do not recognize this all-important tension. They ignore the fact that they are called to something higher than being an animal, that they should allow the spirit to live in them.

No matter at what stage of evolution the creature was that became the first man, God breathed his breath into it and formed it in its inner potentials into an image of God. It has not yet been revealed to us, I believe, what potentials for wonderful experiences are given to man. It must have been an unbelievable religious experience for this first man when, coming from the animal world, he suddenly experienced God – we humans seldom experience God like that, breathing his living breath into us and speaking directly to our hearts. Yet the same thing happened to Mary when she conceived Jesus by the Spirit. So I do not see that any great problem arises between the story of creation in the Bible and theories of evolution.

If man were completely evil and corrupt through and through before being born again of the Holy Spirit, there would be neither room for the divine image nor the possibility of stirring his conscience. We should find an inner vision and an inner reverence for this fact that man is an image of God, seeing with what devotion he opens up to love and with what wonder he submerges himself in streams of holiness. How deeply and wonderfully the human heart can grasp and understand!

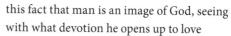

Caspar David Friedrich, *The Wanderer above the Sea of Fog*, oil on canvas, 1818 (detail)

LET US LOOK AT OUR world again: forests, meadows, birds, deer, a valley in the moonlight, a sunset, the starlit heavens. When we think of the starry heavens, the question arises: could God really be so materialistic as to have created so much visible life just here on this little speck of dust we call Earth, leaving everything else absolutely dead? Anyone who thinks deeply will find this impossible; and I am convinced that, just as this earth has an earth-spirit, a prince of this world, so each star has an angel, a spiritual prince or spiritual authority who animates and rules over it. The sun has a fire-angel; the millions of suns in the universe have fire-angels.

We have no knowledge of the beauty of the angel world; very few have seen it. But if we had the possibility of experiencing the star worlds with their angels or spiritual authorities, I think we would be amazed how wonderfully tender God's creation is, with the tenderness of virginity, and yet how wonderfully powerful and manly. ➤

J. Heinrich Arnold (1913–1982) was elder of the Bruderhof and the author of several books including In the Image of God *(Plough, 1977), from which this reading is taken.*

Artwork by Caspar David Friedrich (public domain)

Taquen, *The Past*, acrylic on handwoven fabric

New medical technologies promise
to let us mold our bodies at will.
Should we use them?

The Lords *of* Nature

EDMUND WALDSTEIN

I**N DECEMBER 2019**, the Chinese biophysicist He Jiankui was
sentenced to three years in prison for using CRISPR technology to
edit the genomes of twin babies conceived *in vitro*, to make them
resistant to the HIV virus. His action was widely condemned for its
ethical implications, not least that the risk of editing the children's
genetic code could have unforeseen consequences for their health.
Nevertheless, many experts predict that gene editing will sooner or
later become acceptable practice. In 2020, a panel of experts argued
that while the world "was not yet ready for gene-edited babies,"
approval procedures should be developed for such time that "technical
hurdles were cleared and societal concerns were addressed."[1]

Why is it assumed that "societal concerns" about CRISPR babies
will eventually be addressed? I think the assumption stems from deep
roots in modern culture, in a way that another recent controversy
helps illustrate. In 2015 the Italian surgeon Sergio Canavero gave an

1.   Andrew Joseph, "Expert panel lays out guidelines for germline editing, while warning
     against pursuit of 'CRISPR babies'," *Stat*, November 3, 2020.

---

*Edmund Waldstein, OCist, is a monk of Stift Heiligenkreuz, a Cister-
cian abbey in Austria.*

interview to *Newsweek* in which he spoke of his hopes of developing head transplants for human beings. Head transplants could be used as therapy for many conditions, Canavero said, including gender dysphoria.[2]

Canavero's hopes may remain a fantasy, but they illustrate a typically modern attitude toward human nature, which joins together a desire to dominate nature through human reason with an embrace of subjective feelings of authenticity. This modern synthesis of rationalism and irrationalism has for many people come to seem as matter-of-course as the air we breathe. Yet it is riven by deep contradictions – and contradicts, too, the basic teachings of Christianity.

One way of understanding modernity, indeed, is as a re-interpretation of the Christian doctrine of the lordship of man over the visible creation, including himself. This is a re-interpretation, however, that radically distorts what lordship over nature means. In the Christian tradition, man has only a relative and limited lordship over his body, which he has received from his Creator to be tended in accordance with the purposes and potentials that God has laid into it. But in the modern synthesis of rationalism and irrationalism, man is absolute lord over his body – a body that he may, and perhaps must, change in accord with his authentic inner feelings.

Descartes's Divided Reality

The rationalist half of the modern synthesis can be traced back to René Descartes (1596–1650). A key figure in the development of modern science, Descartes also developed a metaphysical foundation for it. The details of his metaphysics were discarded by later thinkers, but some of its key premises were preserved. The old "scholastic" science had been ordered to the contemplation of the truth for its own sake. But Descartes (following Francis Bacon) thought the new science should be ordered to lordship over nature. Thus Descartes writes:

> [New notions in physics] opened my eyes to the possibility of gaining knowledge which would be very useful in life, and of discovering a practical philosophy which might replace the speculative philosophy taught in the schools. Through this philosophy we could know the power and action of fire, water, air, the stars, the heavens and all the other bodies in our environment, as distinctly as we know the various crafts of our artisans; and we could use this knowledge – as the artisans use theirs – for all the purposes for which it is appropriate, and thus make ourselves, as it were, the lords and masters of nature.[3]

Descartes's system divides reality into two realms: the purely spiritual realm of the "thinking thing" (*res cogitans*), and the purely quantitative realm of the "extended thing" (*res extensa*), that is, the bodily natural world. The human body, as a mere "extended thing," is thus sharply separated from the human soul. In Descartes's words:

> I can infer correctly that my essence consists solely in the fact that I am a thinking thing. It is true that I may have . . . a body that is very closely joined to me. . . . On the one hand I have a clear and distinct idea of myself, in so far as I am simply a thinking, non-extended thing; and on the other hand I have a distinct idea of body, in so far as this is simply an extended, non-thinking

2. Conor Gaffey, "Head Transplants 'Could Replace Gender Reassignment Surgery' for Gender Dysphoria," *Newsweek*, February 27, 2015.

3. René Descartes, *Discourse on the Method*, Part 6, AT 6.61–62, in *The Philosophical Writings of Descartes* 1, trans. John Cottingham et al. (Cambridge University Press, 1985), 142–143.

thing. And accordingly, it is certain that I am really distinct from my body, and can exist without it. [4]

Thus the whole corporeal world, including the human body, is seen by Descartes as neutral material to be dominated by the human soul.

Although later thinkers rejected various parts of the Cartesian system, this thought pattern continues to influence Western culture, as we see in Sergio Canavero. In Canavero the "thinking thing" has come to be identified with the brain, not the soul. Nevertheless, the distinction between "thinking thing" and "extended thing" is preserved. For Canavero human beings *are* their brains, and their bodies can therefore be switched.

The lordship of Descartes's "thinking thing" over the physical world is arbitrary; this lord cannot find purposes intrinsic to natural things that could serve as a guide to the exercise of his lordship. He must himself decide what purposes he will impose on natural things, including his own body. Insofar as his "self" is not a part of nature, he has no reason to pursue "natural" goals. C. S. Lewis points out that this raises a problem:

> The final stage is come when Man by eugenics, by pre-natal conditioning, and by an education and propaganda based on a perfect applied psychology, has obtained full control over himself. Human nature will be the last part of Nature to surrender to Man. [5]

CRISPR technology holds the potential for the kind of thing that Lewis was describing: to control and alter human nature. The scientists who exert such control over a baby's genome

Taquen, *The Self,* acrylic on handwoven fabric

About the artwork: Taquen, an artist based in Madrid, Spain, sees humans as part of nature: "I do not understand a society that is alienated from nature." This series, *Part of a Place*, displays an interaction and mutual respect between the medium and painting, between subject and background. The hemp used for the canvas is grown wild and handwoven in the Nepalese Himalayas, where the artist took inspiration for this series.

are imposing their own ends on him – ends that may or may not be beneficently motivated, but are accountable only to them:

> The Conditioners, then, are to choose what kind of artificial *Tao* [set of ethics or values] they will, for their own good reasons, produce in the Human race. They are the motivators, the creators of motives. But how are they going to be motivated themselves? [6]

Lewis shows that as soon as an objective natural order of purposes, goals, and goods has

---

4. René Descartes, "Meditations on First Philosophy," Meditation 6, AT 78, in *The Philosophical Writings of Descartes* 2, 54.
5. C. S. Lewis, *The Abolition of Man* (MacMillan, 1947), 37.
6. Lewis, *The Abolition of Man*, 39.

Taquen, *Memory*, acrylic on handwoven fabric

cognitive science, as well as for psychologists including B. F. Skinner) there is no "thinking thing": everything is pure extension. Human consciousness is an illusion produced by mechanical reactions of matter.

The Romantics protested against such reductionism: No! There is more to the human spirit than that. They did not, however, understand that "more" as something with a stable nature – in the way that Christians understand the soul, for example – but as a dynamic possibility. Jean-Jacques Rousseau wrote of an "inner voice" of nature. Unlike the tradition going back to Aristotle, he did not understand natural beings as having purposes or ends that must guide their development if they are to thrive. Rather, for him this "inner voice" is a creative principle which itself brings purpose into existence.[8] This idea was further developed by Johann Gottfried Herder and the Romantics, in ways described by the philosopher Charles Taylor:

> Fulfilling my nature means espousing the inner élan, the voice or impulse. And this makes what was hidden manifest for both myself and others. But this manifestation also helps to define what is to be realized. . . . This obviously owes a great deal to Aristotle's idea of nature which actualizes its potential. But there is an importantly different twist. Where Aristotle speaks of the nature of a thing tending towards its complete form, Herder sees growth as the manifestation of an inner power (he speaks of "*Kräfte*"), striving to realize itself externally.[9]

This view is certainly appealing in contrast to the mechanistic alternative, but it is nevertheless wrong. In contrast to Aristotle's

been abandoned, man has no reason to choose *x* rather than *y*. All that remains is arbitrary will based on subjective feelings.

> Everything except the *sic volo, sic jubeo* ["thus I desire, thus I command"] has been explained away. . . . When all that says 'It is good' has been debunked, what says "I want" remains.[7]

Such a completely arbitrary ethic is, however, not satisfying. Hence it is not surprising that pure Cartesian rationalism is generally combined in our times with a quite different ethic: the Romantic ethic of authenticity.

The Romantic Reaction

Romanticism (in a wide sense) began as a reaction against the reductive materialism of the Enlightenment. Many Enlightenment thinkers after Descartes abandoned the spiritual side of his system of thought. For thinkers such as Thomas Hobbes and Julien Offray de la Mettrie (and for contemporary philosophers such as Daniel Dennett, who take on the mantle of

7. Lewis, *The Abolition of Man*, 40.

8. See Charles Taylor, *Sources of the Self: The Making of the Modern Identity* (Harvard University Press, 1989), chapter 20.
9. Taylor, *Sources of the Self*, 374–375.

understanding of nature as having intrinsic meaning and purpose, the Romantic understanding of the inner voice of nature leads to subjectivism: each human person must give authentic expression to his or her particular inner voice. Moral goodness consists not in pursuing the goals of a common human nature, but in creatively developing one's authenticity, a personalized telos which each must discover or create.

There is certainly something of the truth here; we *are* called upon to cooperate in our own formation. Contra the Romantics, though, our natures are not infinitely malleable, to be formed and re-formed by the arbitrary dictates of our will. We learn who we are not through gazing at ourselves, but through our relationships with others, through common experience of human nature, through the development of virtues. We learn how to tend our bodies and honor them through perceiving them as good, as what we are *given*.

The synthesis of the Romantics' celebration of authenticity with Descartes's rationalism finds a fitting if obviously extreme symbol in Canavero's suggestion of head transplant as a treatment for gender dysphoria. According to this way of thinking, which is held by many far more mainstream than Canavero, a man who feels the inner voice telling him that he is a woman must follow it in order to live an authentic life. He must exchange his body for a woman's body. Barring such a radical and clearly impossible solution, he may alter his existing body to become more feminine. His feeling of disjunction is powerful, painful, and real. Although this feeling is quite independent of the natural tendencies of his body, it gives a powerful *motive* for action: a Romantic motive to seek authenticity by way of the Cartesian action of lordship over nature.

Here then is a prime example of Cartesian rationalism becoming the instrument of Romantic irrationalism. The brain plays the role of the "thinking thing," which has absolute lordship, even tyrannical lordship, over the body as extended thing. It does not see the body as part of the self, with its own good nature to be tended and honored – and so it can go so far as to change which body belongs to it.

To be sure, modern medicine changes human bodies all the time, in ways most reasonable people are grateful for – for example, through bone-marrow transplants, cleft-palate surgery, or indeed certain kinds of hormonal treatments. Yet there is a vast difference between changing a body for therapeutic ends – to restore it to its natural function – and changing a body to alter its very nature.

Even in drawing the line against any drastic and irreversible changes, the proper response to persons experiencing dysphoria – persons created in the image of God – is compassion and mutual care within a community in which all are valued and beloved members. There is a tradition older than Cartesian rationalism or Romanticism that offers a more coherent account of the nature of body and soul. This same tradition directs us to bear each other's burdens in a spirit of practical, self-sacrificial love, and commands respect for the dignity of the human person. This is the tradition we turn to next.

Lordship in Christianity

The Christian view diverges sharply from the modern synthesis of Cartesian rationalism and Romanticism. Unlike Romanticism, Christianity understands creation as deeply rational through and through, an expression of the wisdom of the Creator. Yet unlike Cartesian rationalism,

> Christianity understands creation as deeply rational through and through, an expression of the wisdom of the Creator.

Christianity sees the lordship of human beings over creation as legitimate only in relation to the Creator's wisdom: human beings must rule creation in accordance with its deep orderliness, helping natural things achieve the purposes for which they were created.

In the creation story in Genesis, God appoints man as lord over other creatures in God's stead: "Then God said, 'Let us make man in our image, after our likeness; and let them have dominion over the fish of the sea, and over the birds of the air, and over the cattle, and over all the earth, and over every creeping thing that creeps upon the earth'" (Gen. 1:26). Over the centuries, Christian tradition has sought to explore what man's calling to dominion means. Thomas Aquinas, in particular, brings the insights of Aristotelian philosophy and Roman jurisprudence into fruitful dialogue with the biblical account of lordship.

> The ecological movement, to be consistent, must recognize the implications of respect for nature's dignity for our understanding of humanity.

Aquinas expounds his theology of creation with the help of Aristotle's understanding: man is a unity of matter and form, body and soul.[10] Neither is complete in itself; they are only complete in their unity. Further, Aristotle suggests four causes to explain the whys of existence. A statue may be explained in part by its material, "marble," or by its form, "a young man." We might point to the agent who made the statue, "Praxiteles," or to the purpose that the agent had in mind when he shaped the matter into that form: "the honor of the god Hermes."[11] It is this "final cause," the cause of causes, for which a thing exists and without which none of the other causes have effect.[12] The good is the greatest of the causes: first in intention, though often last to come about in reality, as when the statue is set up in the temple.

For Aquinas, the causality of the good is found not only in human action, but in all natural causality. Everything must have an end, or it would neither exist nor act. Every natural thing has a *nature*, a principle of activity ordered to its end. Each thing strives in accordance with its nature for its own completion, and for the kind of activity that is proper to it. A tree grows, stretches out its branches, and photosynthesizes; it does what a tree does. Human beings can understand the end toward which nature orders them and strive consciously to achieve it: they must in a sense cooperate in their own completion. This understanding of the human end is not an invention or a creation, but the discovery of something inscribed in our nature by our Creator. The end he gives us is to do truly human activities well – to act wisely, justly, courageously, generously, and moderately. Aquinas describes this natural tendency toward the good as the *art of our Creator*:

> Nature is nothing other than the reason (*ratio*) of a certain kind of art, namely God's art, impressed upon things, whereby those things are moved to a determinate end. It is as if a shipbuilder were able to give timbers the wherewithal to move themselves to take the form of a ship.[13]

God has absolute lordship over all creation, because he created and sustains all things, and is their first principle and last end. He gives to all creatures their being, their nature, and their ends. He inscribes their purposes within them, and moves them towards those purposes. Creatures, on the other hand, can only have a relative and limited lordship over anything.

10. Thomas Aquinas, *Summa theologiae*, Ia, q.76.
11. Aristotle, *Physics*, 194b-195.
12. Thomas Aquinas, *Summa theologiae*, Ia, q.5, a.2, ad 1.

13. Thomas Aquinas, *In octo libros Physicorum Aristotelis expositio*, Lib. II, lectio 14, n. 8.

Just as God is the universal cause so God is the universal Lord, while creatures are particular lords, whose lordship depends on God's.

Human beings have direct lordship over their own moral actions, since they cause them by their intellect and will. This lordship – freedom – can only be rightly exercised in accord with the prior Lordship of God. But human beings have only an indirect lordship over their own bodies. In the words of the Thomist philosopher Henri Grenier:

> Man has not direct and absolute dominion over his own life and members, but only the guardianship and use of them. For life and body are prerequisites of man's dominion, and are its foundation. Hence they are not subject to man's dominion.[14]

As the guardians of our bodies, the Christian tradition teaches, we must tend them in accord with the purposes that the universal Lord of all has inscribed into them.

I N 2011, POPE BENEDICT XVI gave an address to the German Parliament. The Pope praised the ecological movement which has been active in German politics since the 1970s, celebrating the renewed respect for the natural world, and the concern to protect it from pollution and destruction. This movement, the Pope argued, correctly identified a problem with the way moderns relate to the natural world: we regard it as mere material to exploit for our own ends. This can lead only to destruction; we must instead learn to respect the dignity of nature. He then, however, pointed out a certain inconsistency in the movement – it does not go far enough when it fails to recognize the implications of respect for nature's dignity for our understanding of humanity:

Taquen, *Remnants of a Wild Garden I*, acrylic on handwoven fabric

I would like to underline a point that seems to me to be neglected, today as in the past: there is also an ecology of man. Man too has a nature that he must respect and that he cannot manipulate at will. Man is not merely self-creating freedom. Man does not create himself. He is intellect and will, but he is also nature, and his will is rightly ordered if he respects his nature, listens to it and accepts himself for who he is, as one who did not create himself. In this way, and in no other, is true human freedom fulfilled.[15]

This is of course the idea which Benedict's successor Pope Francis developed more fully in his encyclical *Laudato si'*. True freedom for all human beings, true liberation from all that wrongly constrains us and robs us of our happiness, can only arise from a deep respect for the nature given to us by God. We will only be free when we see both our souls and our bodies as gifts to be developed in accordance with the deep wisdom that makes them what they are. ❧

14. Henri Grenier, *Thomistic Philosophy* 3 (St. Dunstan's University, 1949), 187.

15. Pope Benedict XVI, address to the German Bundestag, September 22, 2011.

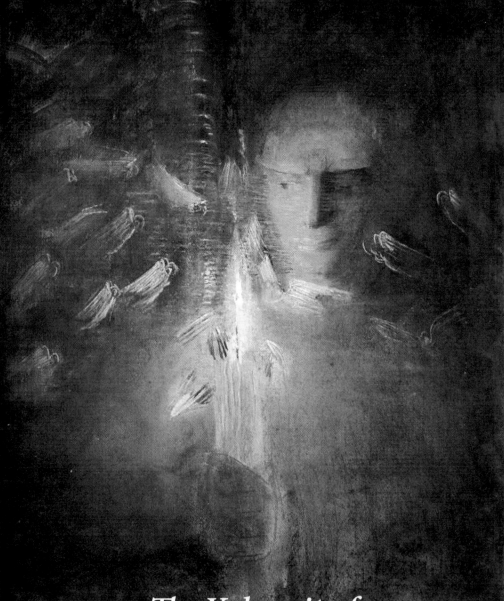

In *The Denial of Death*, Becker says it's in our nature to fear death –
and to transcend that fear through faith.

The Yahrzeit of
Ernest Becker

KELSEY OSGOOD

I WAS TWENTY-SIX YEARS OLD when my grad-school mentor suggested I read *The Denial of Death* by Ernest Becker. I no longer remember exactly what the impetus was – I think it had something to do with a manuscript I was writing, part of which dealt with the urge to pre-memorialize our lives in writing, but it's just as likely she thought it might be helpful to me in my personal life. At the time, I was the sole employee of a true-crime writer who was dying of amyotrophic lateral sclerosis. What had started as a part-time gig taking dictation on subjects like the inner workings of Mafia families had morphed into a situation something between full-time job and enmeshed pseudo-family. I adjusted my boss's pacemaker when it wiggled out of place, fired his visiting nurses for him (their offenses were numerous, and usually cosmetic), sliced his food, operated his ventilator, and when he and his wife picked up and moved to Miami every November, packed my bags and relocated right along with them.

Though I know he treasured me in his own way, my boss was also more than a bit narcissistic – he often described me as an "extension" of himself – and prone to fits of rage that had, in his pre-wheelchair days, regularly turned violent. It is one thing to watch a person die, and another entirely to watch someone so spiritually unsettled, whose ego rests so squarely on an almost parodic machismo, face the end. He seethed and fumed and lashed out at anyone in close range, which meant mostly his wife and, less frequently, me. Young, bereft of colleagues, and guilt-ridden – I was young and he wasn't; I was healthy and he wasn't; I often hated him, despite his agony – I craved a way to process the terrible existential drama that was playing out before me.

Enter Becker. Where Freud saw sex at the heart of human suffering, Becker saw fear of death. Put more broadly, he argues that human consciousness of death – the thing that, he believed, separates us from animals – incites a pervasive, inevitable terror that develops early in childhood. The threat of extinction for creatures so convinced of the vitality of their inner lives is so frightening that we seek to flee from it in all manner of ways, beginning with the development of "character armor," a phrase Becker borrowed from Wilhelm Reich, which essentially translates to personality or lifestyle. As children grow, more options for grappling with this essential knowledge and its attendant fear become available. Some go the comparatively easy route and become "culturally normal," living unexamined lives of low-grade consumerism, social striving, and conformity. Others anesthetize themselves with drugs or alcohol or thoughtless copulation. Some are unable to look away and succumb to madness – I think often of the scene in Ingmar Bergman's *Persona*, when Liv Ullmann's doctor levels with her about her elective mutism: "You think I don't understand? The impossible dream. Not of seeming, but of being."

The paradox is not solvable, of course: preexisting knowledge of our own death is an inevitable part of being human. The trick is to find a way to live without completely ignoring death (that would be a cowardly swim in the shallow end) while also keeping the fear of it at bay for the sake of functionality. Those of us who try to reconcile the dilemma attach ourselves to some "system of heroics" – by aligning ourselves with a seemingly powerful transference object, like a celebrity, demagogue, or strict worldview, we trick ourselves into believing we are impervious to death, in an

Mikalojus Konstantinas Ciurlionis, *Truth*, 1905

Ernest Becker

Kelsey Osgood is the author of How to Disappear Completely: On Modern Anorexia *and has written for the* New York Times, *the* New Yorker, Longreads, *and the* Washington Post. *She is currently working on a book about religious conversion among Millennial and Gen-Z women.*

echo of a child's naïve belief that his parents are immortal and because he is bound to them, he must be too. The cultural scaffolding that accompanies such choices frames useful diversions from the constant terror of extinction, as well: witness the ease with which most humans goosestep when ordered, the vigor with which they wave flags, the care with which they construct their superfan websites – the glee, in other words, they feel when given a mission from a higher source.

There are nobler, more productive, "life-enhancing" illusions, like becoming an artist or having a family, but those too can become frantic, self-defeating bulwarks against extinction if one isn't careful: parents can all too easily view childrearing as a vanity project, while artists have to contend with more isolation and self-doubt than the average person. In fact, Becker posits that nearly the whole enterprise of society is a stage upon which humans play out elaborate, mostly meaningless dramas, all designed to distract from the fact that despite our intellects, our emotions, our quirks of personality, our egos – our bodies, our very *personhoods* – will give out, our flesh will decompose, and we will return to the earth.

Though Becker lingers on all these "immortality projects" he's primarily interested in two systems of heroics: the psychoanalytic and the religious. The fear of death has always existed, but in ancient times religion acted as a near panacea. It accomplished this in a number of ways. First, it acknowledged mystery. Second, it spoke openly of death – even, ingeniously, turning it into a positive in some belief systems, by providing an image of an idyllic afterlife or, in the case of Buddhism, by rebranding it as simply a mode of transport to another life. Religion placed man firmly in the center of the cosmic drama, simultaneously offering him identification with a group. It gave him a vocabulary of symbols and a blueprint for

moral behavior, and, lastly and crucially, gave him a higher entity – God – to devote himself *to*. In short, it was a heroic system that addressed nearly all the enormous and insoluble problems presented by the fact of death.

But by the time *The Denial of Death* was published in 1973, religion had been largely ousted by psychoanalysis, which sought to make men knowable to themselves by means of cold, clinical observation and codification, and which all but guaranteed self-mastery for those who submitted to its ethos.

> The promise of psychology, like all of modern science, was that it would usher in the era of the happiness of man, by showing him how things worked, how one thing caused another. Then, when man knew the causes of things, all he had to do was to take possession of the domain of nature, including his own nature, and his happiness would be assured.

One glance at our world today should prove that this didn't exactly pan out. The problem, Becker said, was that even if man did conquer himself, it would prove a toothless victory, because man's dilemma was centered around the self's inherent destructibility. So while psychoanalysis might be able to treat behavioral symptoms, or to help a person grapple with earthly traumas like the wounds incurred from poor parenting, it could never give people what they really need – a means of reaching outside themselves, toward something profound and inextinguishable, from which they could draw power. Within the boundaries of analysis, man could be a creature, wrestling with his sexual dysfunctions and his behavioral tics, but never a soul whose life has enduring meaning, because psychoanalysis, hell-bent on being a science, refused to couch the human condition in such terms, and had robbed itself of the language required to adequately

Mikalojus Konstantinas Ciurlionis, *Sunset*, 1904

> One by one, Becker shoots down various ways humans alleviate their dilemma as too paltry, too cowardly, too self-destructive.

address it, replacing it instead with jargon. (Of course, this was a conscious choice on the part of Freud, who saw himself as a brave warrior against the forces of "occultism" and all religion generally.)

By the end of the book, it's patently clear to the reader which system wins the battle of the *Weltanschauungen*, even though Becker tries to walk the perimeter around his conclusion instead of straight to its center. He offers some caveats, as well as a few digs: "In some ways [religion] is much worse [than psychoanalysis] because it usually reinforces the parental and social authorities and makes the bind of circumstantial guilt even stronger and more crippling." He tries to hedge by suggesting that different paths should be available, because different people have different needs. (Considering he

just spent the entire book arguing with such poetic force against myriad "failed heroics" and substitute balms, this falls flat.) He tries further to conclude that a whole new heroics system should be created, though he doesn't suggest what it might look like, nor how humans could circumvent the fact that it would be manmade and thus inherently unreliable. Despite these protests, his preference for religion is plastered all over the text. All the thinkers he lauds and quotes liberally – Kierkegaard, William James, Otto Rank – extoll its value. One by one, Becker shoots down various ways humans alleviate their dilemma as too paltry, too cowardly, too self-destructive. At one point, he suggests that psychology could possibly work if it became more of a "lived experience," developed a spiritual vocabulary, or if the therapist acted

> When I placed my existential depression within the context of universal creation and destruction, it became painfully clear that the tools I had been operating with previously were way too small for the job.

Mikalojus Konstantinas Ciurlionis, *Night*, 1904

more as a guru than a scientist. In other words, it could work, but only *if it were structured more like religion*. (Indeed, Becker writes in numerous places that his work as a whole is about a fusion of science with religion, although curiously one almost always seems to be paramount.) Over and over, Becker says that people need something beyond themselves, that exists completely independently, some entity that gives credence to both the body that will decay and the spirit that will endure. There is only one thing that fits that description, and that is religion.

It's fair to say that what Becker is really getting at is not that religion is the best system for processing our relationship to mortality, but that the divine reality religion points to is the only answer that satisfies. He wasn't wrong when he said that religion can become

a calcified, self-satisfying loop just as readily as manmade ideologies, and that religious people can latch on to heroics just as thoughtlessly as an average man fashions himself in the image of the masses or the private follows the orders of his sergeant. Religion can fan the flames of xenophobia and incite war, something Becker found repugnant. Without faith in God underpinning them, certain religious practices just mimic neurotic behaviors and encourage alienation from the body, serving to do little more than torture the believer. But a belief grounded in expansiveness, love, and reverence, that fosters both humility and self-respect, that fixes man's eyes on the stars and his heart – his poor, fragile, doomed human heart – *that* can never fail. "Men should wait [for redemption, or understanding] while using their best intelligence and effort to secure their adaptation and survival," Becker summarizes. "Ideally they would wait in a condition of openness toward miracle and mystery, in the lived truth of creation, which would make it easier both to survive and to be redeemed because men would be less driven to undo themselves and would be more like the image that pleases their Creator: awe-filled creatures trying to live in harmony with the rest of creation." Certainly in our era there is no means to discuss, let alone experience, the "miracle and mystery" of life, death, and faith other than religion, imperfect vehicle for these sacred concepts that it can be.

The Denial of Death came at exactly the right moment in my life, not only because of the circumstances of my employment, but also because by my mid-twenties, I had already tried a number of "heroics systems" – rational atheism, nihilism, mental illness, analysis as "lived experience," writing – and, having failed to find the succor I so craved, was rather sheepishly tiptoeing towards faith. ("Sheepishly" because I had argued so vociferously against it in my youth, and because it wasn't exactly a popular choice for an overeducated young person from a secular background.) When I placed my existential depression within the context of universal creation and destruction, as Becker pushed me to do, it became painfully clear that the tools I had been operating with previously were way too small for the job – I was trying to fix a burst pipe with a pair of tweezers. This was still years before I became a religious Jew and encountered the texts my tradition reveres, which insist that man meditate on the fate of his body.

"Akavya ben Mahalalel said: 'Reflect on three things and you will avoid transgression: Know where you came from, where you are going, and before whom you will have to give an account and reckoning,'" reads a portion of *Pirkei Avot*, a compendium of rabbinic ethical

teachings. "'Where you came from' – from a putrid drop. 'Where you are going' . . . to a place of dust, worms, and maggots." Though I am not Hasidic, I discovered one of my favorite teachings on the human condition in that mystical tradition: *It was said of Reb Simcha Bunem that he carried two slips of paper, one in each pocket. On one he wrote: Bishvili nivra ha-olam – "for my sake the world was created." On the other he wrote: V'anokhi afar v'efer – "I am but dust and ashes." He would take out each slip of paper as necessary, as a reminder to himself.* I found this anecdote so moving that my husband had a necklace made out of a small disc for me, each side engraved with one of Reb Bunem's dicta (Hebrew speakers are sometimes surprised to read it when the "dust and ashes" side faces outward).

My first encounter with Becker was also around the time when the zeitgeist was filled with new calls for public attention to death. The mid-aughts began an age of talking about the "good death," which continues. It's a diffuse movement marked by hipster morticians and *cafés mortels*, themed Instagram accounts that churn out endless content for the grief-stricken, the proliferation of death doulas and end-of-life activists, and books by Atul Gawande, Lydia Dugdale, and others who call for greater awareness and acceptance of death. Some of these efforts Becker would undoubtedly have admired; others I suspect he would have thought counterproductive, in that they propagate the idea that death can be controlled and sanitized, made cheerful, memeable, or painless, both for the dying and the grieving parties. Surely we haven't matured that much as a species – we still mostly relegate our dying to hospitals and institutions. Perhaps not coincidentally, this was also when technocrats supercharged their quest to eliminate death by designing experimental treatments reminiscent of the wildest plot points in science fiction.

There were new and refreshed ventures toward physical perfection, distorting our faces and bodies to look shinier and younger against the reality of bodily decay. Becker would undoubtedly decry these practices. Fearing death is natural and inevitable; attempting to outrun or hack it is spineless.

In the years after I read *The Denial of Death*, I occasionally sought out biographical information on Becker, and was perplexed to find that none of it addressed his religious life. His Wikipedia page simply states that he was born into a Jewish family; the biography on the website of an eponymous foundation, established by an acolyte after Becker's death, primarily covers his life in academia, and mostly focuses on the ways in which religion can be its own failed heroics system, a possibility that comes up quite rarely in the actual book. How could the person who wrote the sentence "the urge to cosmic heroism . . . is sacred and mysterious and not to be neatly ordered and rationalized by science and secularism" not have had a relationship with the divine?

But finally, after years of searching, I tracked down the interview philosopher Sam Keen conducted with Becker on his deathbed – he succumbed to colon cancer at forty-nine. In it, Becker speaks through what must be enormous pain about the vicissitudes of his faith. He tells Keen that although he was an atheist for many years, the birth of his first child jolted him into belief in God – "seeing something pop in from the void and seeing how magnificent it was, unexpected, how much beyond our powers and our ken" – and that he felt the only valuable conversation man can have is a "vertical" one: "I think a person must address himself to God rather than to the future of mankind." Though death hadn't made him more religious (he insisted this was a result of shedding his character armor), his words are those of a man at peace with the idea of returning to a place of dust, worms, and maggots.

I would say that the most important thing is to know that beyond the absurdity of one's life, beyond the human viewpoint, beyond what is happening to us, there is the fact of the tremendous creative energies of the cosmos that are using us for some purposes we don't know. To be used for divine purposes, however we may be misused, this is the thing that consoles . . . I think one does, or should try to, just hand over one's life, the meaning of it, the value of it, the end of it.

It is one of the loveliest and most articulate expressions of faith I've ever encountered.

This past summer, my father-in-law died unexpectedly, while biking on a public trail near his home in Utah. My husband and I forced face shields onto our two toddlers and raced through the airport to make it in time for the funeral, where, according to Jewish custom, we shoveled the dirt directly onto the coffin ourselves. Then we sat in my mother-in-law's home and cried and prayed together, with the help of *shiva* visitors over Zoom. Our nonreligious friends and relatives seemed tense, often shying away from mentioning the loss. But our faith community jumped into gear immediately, bringing us food, helping us ensure my husband could say *kaddish* with the necessary quorum, sharing memories of my father-in-law even if they'd only met him briefly and years ago. Nobody told us he had transcended to a better place, or that we should feel anything other than sad, or even angry with God, as arguing with him is a time-honored Jewish tradition and right. ("*Baruch Dayan Ha'emet*," we Jews say when we first hear someone has passed away. "Blessed is the True Judge," whose judgment we must accept, but are not obliged to enjoy.) The death was shocking and somewhat premature – he

Mikalojus Konstantinas Ciurlionis, *My Road II*, 1907

wasn't a young man, but he was in good health – and not what proponents of an ideal death might call "dignified": he died alone, of undetermined causes, without someone massaging his feet, or his family flanking his bedside, or his signed and notarized health directives laid out on the bedside table. But if you asked me, then or now, whether I thought he would have passed in anger, I would say no. He was a man of tremendous self-awareness and independence, a father of three, a tireless philanthropist who founded an initiative to help refugees settle in Utah, a proud and committed Jew, a romantic who liked to say that even after forty years of marriage, he still dreamed of his bride. Though I still deeply mourn, I know he was a man who knew where he had come from, and where he was going.

Every year since my conversion, I've thought of Ernest Becker on March 6, the day of his death. For Jews, this is called a Yahrzeit, and it's marked by lighting a candle and saying a prayer for the deceased, both things I've considered doing for him, a person I never met but who profoundly altered my view of human nature, the world, and myself. Sometimes I wonder if Becker would think me foolish to want to commemorate him in this way. Surely the person truly liberated from the fear of death would have no need to be remembered or to have little altars erected. But Becker didn't expect us to be pure spirit, just as he didn't think we were only animals. No: he knew very well that we are both. ➤

Editors' Picks

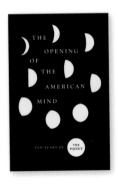

The Opening of the American Mind: Ten Years of *The Point*

(University of Chicago Press)

There's very little surprise in writing today. From the first sentences of most articles, you can tell which of a rather limited collection of categories the piece will fall into: wokeism or anti-wokeism, Trumpism or Resistance, generic secularism or uncritical progressivism or defensive traditionalism.

The Opening of the American Mind, a collection of essays from the first decade of *The Point* magazine, is a welcome exception. In these explorations of contemporary politics and society, one has the rare and thrilling experience of being led word by word on a journey through ideas – a journey where the destination doesn't matter as much as the accompaniment, the participation in a process of understanding that is part of what it means to be human.

Some of the anthology's strongest entries come in the first section, on "the end of the end of history" in the wake of the Great Recession. Etay Zwick's portrait of Wall Street culture and its downstream effects manages to describe high finance evocatively as the realm of "barbarians" without lapsing into clichéd contempt. (Yes, there is room for compassion even here.) Ben Jeffrey's review of Houellebecq's oeuvre recognizes and doesn't look away from the void of meaning in our secular-materialist culture. Later, Jon Baskin casts a critical eye at the idea-industrial complex on the left (and, implicitly, the right) and finds a culture of

the instrumentalization of ideas – and thus of our very nature as thinking beings.

And so, as important as the journey is, the destination is not irrelevant. In their introduction, editors Jon Baskin and Anastasia Berg insist that their project is not relativistic, but rather an attempt to recognize and to nurture the kind of humane and fertile pluralism from which a truth-loving people emerges:

> Pluralism does not release the individual from the responsibility to choose or to judge, because it does not assert that all choices are equal, or that judgments of good and bad are futile. It does demand that we make our choices and judgments with humility, in recognition of the fact that other choices and judgments are both possible and defensible.

In this, the editors articulate a traditional liberalism that evokes the teaching of Pope Francis in his recent encyclical letter *Fratelli tutti*, on the universal brotherhood of man. The editors write that one of *The Point*'s deepest intentions is "that our social and political commitments are strengthened, rather than compromised, by genuine dialogue." Francis writes:

> Authentic social dialogue involves the ability to respect the other's point of view and to admit that it may include legitimate convictions and concerns. . . . When individuals or groups are consistent in their thinking, defend their values and convictions, and develop their arguments, this surely benefits society.

I have been, and remain, skeptical that secular liberalism provides a suitable substrate for the kind of dialogue that forms and sustains communities. But *The Opening of the American*

Mind, along with the continued thriving of *The Point* magazine, provides a hopeful hint of the possibility for renewal – a renewal based on each of us recognizing in others a potential brother or sister in truth.

—*Brandon McGinley,*
Plough *Contributing Editor*

Klara and the Sun: A Novel

*Kazuo Ishiguro
(Knopf)*
We live at the mercy of forces we cannot control. Increasingly, one of those forces is technology. Following his devastating romantic tragedy *Never Let Me Go,* Kazuo Ishiguro's new novel *Klara and the Sun* inhabits a similar dystopian future, and yet the forces he explores are not so much those of technology but rather those of love, faith, and human nature.

The novel takes place in a not-too-distant future of polluted cities, human workforces rendered leisured and useless by machines, children looked after by robots, the looming prospect of singularity. Yet at the heart of this story is not horror but wonder. This is made possible by the novel's narrator, Klara, a robotic AF (artificial friend) who looks after a girl (Josie) whose health is suffering after a procedure of genetic editing (lifting) to improve her IQ. Klara attends to the world with unbridled curiosity, and any knowledge gained serves her primary purpose: to care for a child. Klara notices everything, and speaks about it with simple poeticism, granting a surprising new perspective on life.

The world is enchanted for Klara, full of mysteries she hopes to understand. She is solar-powered, and speaks about the sun as a person of faith would speak about God: with reverence, awe, petition. She notices the kindnesses of the sun when others do not: the way it glances off the shoulders of reunited lovers, pours down upon happy children, and heals homeless men. She trusts the sun implicitly, desiring his "special nourishment" for herself and those she loves, a theme which grows in intensity as Klara begins to petition the sun for Josie's life. At moments, the story feels like it is building to a disappointment, a radical rejection of faith, and yet epiphanic scenes break like a sunrise over the darkest corners of the narrative.

Programmed to care for children, Klara's central preoccupation is human nature. Bent on improving her service, she observes with interest and quiescence the complicated ways of the humans around her. Her most robotic feature is the way she splits contradictory facial expressions into small boxes so she can attempt to understand them. This capacity reveals to her how complex human love can be, one face revealing sadistic pleasure, sadness, and desire all at once. Through Klara's eyes we see that the most inescapable quality of human nature seems to be the way we make ourselves miserable with the internal wars we wage between love and need, gift and resentment, cruelty and kindness. Strikingly, as a robot, Klara often acts more lovingly, more humanely, than the humans she serves.

Like the old fairy tales about mermaids, seeing through the eyes of a creature that is almost human can grant the reader greater clarity about what it means to be human. Klara lives in a world that has given up on humanity and on faith, and yet through her mechanical eyes, we see a world sustained by faith and permeated by love. One character asks Klara, "Do you believe in the human heart? I don't mean simply the organ, obviously. I'm speaking in the poetic sense. . . . Do you think there is such a thing? Something that makes each of us special and individual?" We cannot help but answer yes.

—*Joy Clarkson,* Plough *Contributing Editor*

Writing in the Sand

You can become complacent in faith – and also in doubt.
But God is in the world, just as poetry is in a poem.

CHRISTIAN WIMAN

A FEW YEARS AGO I was asked to give the convocation address at Yale Divinity School, where I have taught for the past decade. Not only did I happen to be reading George Marsden's biography of the great eighteen-century minister and theologian Jonathan Edwards, who was both a student and tutor at Yale, but I happened to have paused at precisely the moment when Edwards himself was about to address the student body. Teaching in an institution to which I would not have been admitted as a student (bad grades, bad "life choices"), I was flattered by the association, and it occurred to me that many of the students in attendance might be as well. To be welcomed into a place with so much august history, so much intellectual curiosity and attainment, so many great names – surely it's worth a moment of pride. But maybe just a moment. In the last chapter of Marsden's book I came across a quote from Ezra Stiles, who was president of Yale when Edwards died. "In another generation," said old Ezra, the works of Jonathan Edwards "will pass into as transient notice perhaps scarce above oblivion, and when posterity occasionally comes across them in the rubbish of libraries, the rare characters who may read and be pleased with them will be looked upon as singular and whimsical."

The pride of accomplishment, the humility of being you. The glory of the door, the reality of the room. They are ancient themes.

There is a profound tension in the eighteenth century between divine grace and human reason. Grace is absolutely beyond all human capacity, one thinker after another (like Edwards) will say, and then they begin furiously reasoning their way toward it. The idolatry of logic – and it does often seem like that – led to some miraculous discoveries, such as those of Isaac Newton. And it led to some pretty strange Nostradamus-like noodlings in the margins of the Book of Revelation as one thinker after another tried to pin down the exact instant of the Rapture. Newton himself engaged in this activity, as did Edwards. In preparing for that talk at Yale, in fact, Edwards read repeatedly the verse from Revelation 16 in which "there fell upon men a great hail out of heaven, every stone about the weight of a talent" and told himself that the hail, in this instance, would be his own remorseless rhetoric, which he would unleash (in Latin) upon any Antichrist (that is, Anglican) who happened to be in the audience.

I have no hailstones. I have no Latin and I have no answers. What I do have instead are two things. The first is a first-century Jew from Nazareth well known for his oratorical skills but nevertheless, at a crucial moment in his ministry, remaining silent and writing in the sand. It is a strange moment – and one of my very favorite stories from the New Testament. I'll come back to that. The second thing is another form of writing in the sand: poetry.

THE PLACE WHERE WE ARE RIGHT
Yehuda Amichai

From the place where we are right
flowers will never grow
in the spring.

The place where we are right
is hard and trampled
like a yard.

But doubt and loves
dig up the world
like a mole, a plow.
And a whisper will be heard in the place
where the ruined
house once stood.

A whisper will be heard in the place where the ruined house once stood. By which Amichai means, I think, that even though our human pride might wreak havoc upon our houses, there might, if we have the proper humility, arise a living whisper out of the ashes, something resuscitating and revitalizing, something close, perhaps, to a still, small voice.

Now, the circles through which I move, even the religious ones, constitute a pretty "safe space" for this poem. It requires no great courage for me to celebrate its spirit of productive doubt. But I must admit, I do hear

Jonah Calinawan, *The River Crossing*, cyanotype print with digital drawing

Christian Wiman is the author of numerous books of poetry, prose, and poetry in translation. Survival Is a Style (Farrar, Straus and Giroux, 2020) is his most recent collection.

the skeletal chuckle of Jonathan Edwards in my own mind – his ambition, after all, was to be "God's trumpet." And you can make an idolatry of doubt. You can become so comfortable with God's absence and distance that eventually your own unknowingness gives you a big fat apophatic hug. One could argue that when doubt becomes the path of least resistance it becomes the very thing that a faithful person must most resist. And resistance is often a matter of language.

BACKWARD MIRACLE
Kay Ryan

Every once in a while
we need a
backward miracle
that will strip language,
make it hold for
a minute: just the
vessel with the
wine in it—
a sacramental
refusal to multiply,
reclaiming the
single loaf
and the single
fish thereby.

Initially, it might seem that Kay Ryan's poem falls more on the "doubt" side of the ledger than on the "faith" side. The great stumbling block for modern consciousness with regard to the Gospels are the miracle stories. Why should we believe that the laws of reality, which seem so implacably inflexible for us, were mysteriously suspended for a few years in first-century Palestine? Indeed, there has been during the past century or more a whole theological movement to "demythologize" the Gospels. At its best, such thinking has helped to re-sacralize matter and restore primary importance to immediate existence (human and otherwise). This is obviously Jesus' intention throughout the Gospels, even when – actually, especially when – he is performing miracles. At times, though, demythologizing the Gospels has led faith to take refuge in neutered and confused clichés. How many sermons, how many blurbs on the backs of poetry books, praise the capacity for "discovering the extraordinary in the ordinary"? For a thought to become this common is no guarantee it's rotten, but one might want to give it a good sniff.

Ryan's poem raises this whole question while inverting the terms. The miraculous is so "common" (as in beneath us) that sometimes we need to be jolted back by and to the particular. Don't be too quick to transcend, her poem tells us. Being precedes meaning.

And yet, we need this backward miracle only "every once in a while." Why? Because if it can become too easy to transcend, it can also become too easy not to. One can become so disenchanted, so adapted to the "reality" of one's immediate senses and experience, that reality itself, which surely is stranger than our minds can circumscribe, becomes pinched, partial, even inert. (Attention *catalyzes* existence: "The eye with which I see God," says Meister Eckhart, "is the eye with which God sees me.") Ryan's poem, in essence, unsays itself: don't be too quick to *eschew* transcendence. Don't be too sure that being is not filled with meanings that are the task of one's life to discern.

Being and meaning: two ways in which the mind relates to – or, in the case of the former, participates in, even fuses with – life. Christ and Jesus: two names that are the source and pattern for that way of relation. Christ is Being itself. Jesus is one specific meaning that Being acquired at one specific date in history (and forever after). And they – being and meaning, Christ and Jesus – are one thing.

They say unto him, Master, this woman was taken in adultery, in the very act. Now Moses in the law commanded us, that such should be stoned: but what sayest thou? . . . But Jesus stooped down, and with his finger wrote on the ground, as though he heard them not.   John 8:3–11

Now consider the moment when Jesus writes in the sand in John 8. The Pharisees have come to him in the temple courts with a woman accused of adultery. They ask what they should do with her, given that Mosaic law demanded her death. This is a trap. If he says "stone her," then he's breaking Roman law. If he says "don't stone her," then he's setting himself above Jewish law – either way, he's in trouble. Instead of answering, Jesus bends down and writes in the sand. When they persist – in outrage, one can safely assume, because think how irritating that would be – he says his famous line, "Let him who is without sin cast the first stone." Then he bends down and writes in the sand again.

What does he write? That's the first place the mind goes, isn't it? It's certainly the direction a lot of sermons take. (Though some scholars question whether Jesus could write at all; I'm not equipped to weigh in on this with any authority, but if he couldn't write, what was he doing down there, doodling?) Some say he was writing down the names of those self-righteous and accusatory Pharisees standing around him and the woman. Some say it was their sins, others that he was writing down specific verses from the Old Testament. To this layperson all three seem about as likely – and as consistent with Jesus' character – as doodling. So how to read this passage, which by the way is probably not even part of the original Gospel, since one thing that scholars *do* agree on is that this anecdote of Jesus writing in the sand was added later.

This is a job for a poet. Or, more accurately, a job for those who know how to read poetry, because this scene operates as a kind of poem. It is meant to be experienced, not dissected or "filled in." It is suspended between the metaphorical and the literal, between myth and witness.

Consider the mythic elements. First, there's the act itself, of writing on the ground surrounded by inquisitors. Who does that? Just try it the next time you find yourself in a heated meeting. Then, too, Jesus writes with his finger, not an implement of some kind. The Word (capital W) inscribes the word (lower case) upon reality itself – reenacting, I would argue, and perhaps salvaging, that original moment when the Word of God became the word of man. Also, it is metaphorically suggestive that Jesus writes on the earth, not on a tablet, as if the law had come alive, as if the closed world of human religion represented by the Pharisees had been blown open and shown to be as transient and perishable, but also as immediate and meaningful, as this glorious earth that is all around us.

On the other hand: consider the documentary details of the scene. We are told exactly how the crowd is arranged and the order in which the Pharisees depart. We know the woman is not simply being accused of adultery but has been "caught in the act." Then there's the fact that this act of writing in the sand really *does* feel like something the unlikely, unpredictable, and often decidedly unhuggable Jesus would do. It has the feel of a witnessed event, whether or not it was.

Marianne Moore famously described the successful poem as containing "imaginary gardens with real toads in them." The phrase is apt for this scene from John as well. It is one of those moments we come to again and again in the Gospels – whether it's a parable whose message is either implacably opaque or so transparently obvious that it amounts to its own kind of koan, or the silence after Pilate's question "What is truth?" which you can still hear two millennia later, or in this moment when Jesus "writes" something that you will never read, never "understand," and thus maybe, just maybe, never forget.

Does this mean that religion and poetry are essentially the same thing? Best to let a poem provide an "answer."

POETRY AND RELIGION
Les Murray

Religions are poems. They concert
our daylight and dreaming mind, our
emotions, instinct, breath and native gesture

into the only whole thinking: poetry.
Nothing's said till it's dreamed out in words
and nothing's true that figures in words only.

A poem, compared with an arrayed religion,
may be like a soldier's one short marriage night
to die and live by. But that is a small religion.

Full religion is the large poem in loving repetition;
like any poem, it must be inexhaustible and complete
with turns where we ask Now why did the poet do that?

You can't pray a lie, said Huckleberry Finn;
you can't poe one either. It is the same mirror:
mobile, glancing, we call it poetry,

fixed centrally, we call it religion,
and God is the poetry caught in any religion,
caught, not imprisoned. Caught as in a mirror

that he attracted, being in the world as poetry
is in the poem, a law against its closure.
There'll always be religion around while there is poetry

or a lack of it. Both are given, and intermittent,
as the action of those birds—
crested pigeon, rosella parrot—
who fly with wings shut, then beating, and again shut.

God is the poetry caught in any religion, a law against its closure. God is in the world as poetry is in the poem, a law against its closure. "Have you felt so proud to get at the meaning of poems," Walt Whitman famously asks in "Song of Myself," by which he means, of course, that that very pride of understanding is just another form of ignorance, and ignorance is not at all the same as a fractious and catalyzing – as opposed to a cozy and complacent – unknowingness. Have you felt proud to know the meaning of scripture, the right kind of theology, to know what you believe, or even, perhaps, that you don't believe in anything at all? Perhaps you have forgotten the law against closure. By all means, let us declare our faith, if we have any; let us be "God's trumpets." Because in that first Amichai poem it is not only "doubts" that dig up the yard and restore the ruined house, but "loves" as well, and love is decidedly active and declares itself. But let us also keep in mind the ineluctable law against closure, the poetry of reality.

PLOUGH BOOKLIST

Recent Releases

Pillars: How Muslim Friends Led Me Closer to Jesus
Rachel Pieh Jones
Foreword by Abdi Nor Iftin

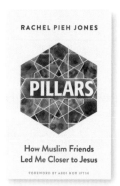

When Rachel Pieh Jones moved from Minnesota to rural Somalia with her husband and twin toddlers eighteen years ago, she was secure in a faith that defined who was right and who was wrong, who was saved and who needed saving. But what happens when one's ideas about God and the Bible crumble and the only people around are Muslims?

Marilyn R. Gardner, author, *Between Worlds:* As an American raised in a Muslim country, I have waited for a book like *Pillars* all my adult life, a personal book that discovers similarities and honors differences between Christianity and Islam, a book that, pillar by pillar, builds bridges of greater understanding across what are often chasms of disconnect.

Softcover, 280 pages, ~~$17.99~~ $10.79 with subscriber discount

Thunder in the Soul: To Be Known by God
Abraham Joshua Heschel
Edited by Robert Erlewine, introduction by Susannah Heschel

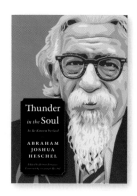

Like the Hebrew prophets before him, the great American rabbi and civil rights leader reveals God's concern for this world and each of us. Only by rediscovering wonder and awe before mysteries that transcend knowledge can we hope to find God again. This God, Heschel says, is not distant but passionately concerned about our lives and human affairs, and asks something of us in return.

Publishers Weekly: Illuminating. . . . Those new to Heschel will appreciate this accessible introduction.

Softcover, 168 pages, ~~$12.00~~ $7.20 with subscriber discount

The Living Word: Inner Land – A Guide into the Heart of the Gospel, Volume 5
Eberhard Arnold

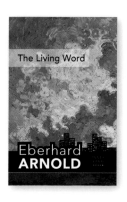

The living word, Eberhard Arnold writes, is greater than the words of the Bible, which even the devil used to tempt Jesus. But when the Holy Spirit speaks this living word into the hearts of those who have set out on the way of discipleship to Christ, the deepest meaning of the scriptures are opened up to them.

Chris Faatz, Powell's Books: *Inner Land* is a bold and challenging invitation to the path of discipleship that speaks to both the terrors and the hopes of our time. Along with the likes of John Woolman, Thomas Kelly, and Dorothy Day, Eberhard Arnold is one of the great secrets of radical Christianity.

Hardcover, 96 pages, ~~$20.00~~ $12.00 with subscriber discount

Children in Nature

When Spring Comes to the DMZ
Uk-Bae Lee

Batchelder Honor Winner, 2020 ALA Youth Media Awards

Korea's demilitarized zone has become an amazing accidental nature preserve that gives hope for a brighter future for a divided land. This unique picture book invites young readers to a place where salmon, spotted seals, and mountain goats freely follow the seasons and raise their families in this 2.5-mile-wide, 150-mile-long corridor where no human may tread. Uk-Bae Lee's lively paintings plant the dream of a peaceful world without war and barriers, where separated families meet again.

Kirkus Reviews, Starred Review: This bittersweet picture book walks through the four seasons at Korea's heavily weaponized demilitarized zone, celebrating the nature that thrives there while mourning the human cost of this border wall.

Hardcover picture book, 40 pages, ages 4-8, ~~$17.95~~ $10.77 with subscriber discount

Charlie the Tramp
Russell and Lillian Hoban

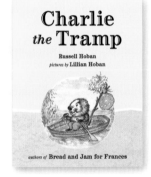

Charlie the Beaver wants to be a tramp when he grows up. "Tramps don't have to learn how to chop down trees and how to roll logs and how to build dams. Tramps just tramp around and have a good time." Charlie sets off with his bundle. But when he hears water trickling, he can't get to sleep. Will he be able to resist the urge to make it stop?

The Atlantic: An especially memorable character study of a runaway beaver . . . unique, humorous, precise of speech.

Hardcover, 48 pages, ages 3-8, ~~$16.00~~ $9.60 with subscriber discount

Their Name Is Today: Reclaiming Childhood in a Hostile World
Johann Christoph Arnold
Foreword by Mark Shriver

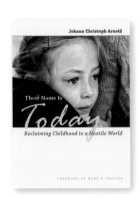

Despite a perfect storm of hostile forces that are robbing children of a healthy childhood, courageous parents and teachers who know what's best for children are turning the tide. Arnold takes on technology, standardized testing, overstimulation, academic pressure, marketing to children, over-diagnosis and much more, calling on everyone who loves children to combat these threats to childhood and find creative ways to help children flourish.

Jonathan Kozol: Beautiful. . . . It is Arnold's reverence for children that I love.

Paperback, 189 pages, ~~$14.00~~ $8.40 with subscriber discount

Poets in Nature

Water at the Roots: Poems and Insights of a Visionary Farmer
Philip Britts
Foreword by David Kline, edited by Jennifer Harries

A farmer, poet, activist, pastor, and mystic, Britts (1917–1949) has been called a British Wendell Berry. His story is no romantic agrarian elegy, but a life lived in the thick of history. As his country plunged into World War II, he joined the Bruderhof, an international pacifist community, and was soon forced to leave Europe for South America. Amidst these great upheavals, his response – to root himself in faith, to dedicate himself to building community, to restore the land he farmed, and to use his gift with words to turn people from their madness – speaks forcefully into our time.

Norman Wirzba, author, *Food and Faith*: For those seeking a healthy and peaceful world, this book will be a provocation to a better way of living.

Paperback, 179 pages, ~~$16.00~~ $9.60 with subscriber discount

The Heart's Necessities: Life in Poetry
Jane Tyson Clement
Introduction by Becca Stevens, edited by Veery Huleatt

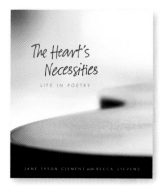

What are the heart's necessities? It's a question Jane Tyson Clement asked herself over and over, both in her poetry and in the way she lived. Her observation of the seasons of the soul and of the natural world have made her poems beloved to many readers, most recently jazz artist Becca Stevens. Clement's poetry has gained new life – and a new audience – as lyrics in the songs of this pioneering musician of another century.

Sarah Ruden, author, *The Face of Water*: Through hard-won religious commitment, Jane Tyson Clement's poems rose from feminine eloquence, in the manner of Edna St. Vincent Millay and Anne Morrow Lindbergh, to something closer to universal.

Paperback, 160 pages, ~~$19.95~~ $11.97 with subscriber discount

The Gospel in Gerard Manley Hopkins
Edited by Margaret R. Ellsberg, foreword by Dana Gioia

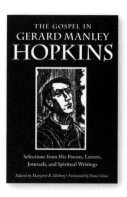

We love Hopkins not only for his literary genius but for the hard-won faith that finds expression in his verse. Who else has captured the thunderous voice of God and the grandeur of his creation on the written page as Hopkins has? Seamlessly weaving together selections from Hopkins's poems, letters, journals, and sermons, Ellsberg lets the poet tell the story of a life-long struggle with faith that gave birth to some of the best poetry of all time.

Hopkins Quarterly: Ellsberg is effective in her contagious love of Hopkins' poetry. She is a true Hopkins believer and faithfully discharges her devotion to the poet.

Paperback, 268 pages, ~~$18.00~~ $10.80 with subscriber discount

creating malleable cases along the edge. Patrick Tomassi's article, while not exactly sympathetic to the antifa and Proud Boy activists it profiles, makes clear that they are entirely sincere in seeing themselves as opponents of the True Violence (of corrupt systems, of false narrative, of entrenched power) despite the fact that they superficially appear to be the ones carrying truncheons and spoiling for a fight. In their own way, these militants are more scrupulous about definitions than the much larger internet world of armchair activists. They seem willing to narrowly define their enemies and honestly confront them in a way that acknowledges a difference in kind between a physical threat and the abstract "harm" of heated disagreement.

At the same time, the problem of how to understand violence in virtual spaces, and how it overlaps with the real physical violence it occasionally inspires, remains mostly unresolved. The Sermon on the Mount instructs that calling a man a "fool" is akin to murder, and that seems to be relevant in an era where every act of real violence seems to rest on the tip of an iceberg of a thousand angry words. Indeed, the sharp delineation of the boundary between physical and verbal violence feels constantly gamed and exploited by the sort of activists described by Tomassi, who relentlessly goad their opponents with mocking chants and jeers into

striking the first blow. While few of us are literal participants, this strategy of provoking our enemies is a far more ubiquitous pattern of sin, easily seen in the temptation to win debates by tempting ideological adversaries to lose composure and escalate rhetoric. In the same way that a culture of pornography and the instrumentalization of sexual behavior has complicated the Christian obligation to avoid sins of lust, the engagement-maximizing design of social media has created an environment that requires heroic virtue to resist wrath. Perhaps we need to look harder at our complicity in creating and sustaining the malign aspects of that environment.

Edward Hamilton, Longview, Texas

THE WHITE ROSE

On Andrea Gross Ciponte's "Freiheit!":

Stories of martyrs require care. We might valorize suffering, or try to become famous through persecution, rather than making true sacrifice for the good, including through hidden faithfulness. We might avoid the testimony of the martyrs altogether, turning aside from their discomforting questions about complacency towards God and casual disregard of our neighbor. Andrea Grosso Ciponte's *Freiheit!: The White Rose Graphic Novel* handles the stories of Hans and Sophie Scholl and the anti-Nazi resistance movement The White Rose with both care and verve. I warmly commend the book.

However, I do wonder if readers forming their first impression from *Freiheit!* will understand how their Christian faith informed the resisters' activities. An excellent companion piece to *Freiheit!* might be the source collection *At the Heart of the White Rose*, also published by *Plough*, which contains key letters and diary entries from Hans and Sophie Scholl. Readers of *Freiheit!* might profit from the tender and searing honesty of Sophie's prayers: "My soul is like an arid waste when I try to pray to you, conscious of nothing but its own barrenness. My God, transform that ground into good soil so that your seed doesn't fall on it in vain."

Alongside its beautiful illustrations and storytelling, *Freiheit!* helpfully includes the full text of all six White Rose leaflets, which appealed on theological grounds to Germans to recognize and resist the evils of fascism and the mistreatment of vulnerable human beings; their words leap off the page: "We will not be silent . . . we are your bad conscience, the White Rose will not leave you in peace!"

Joshua Heavin, Dallas, Texas

Midwestern Logistical Small Talk

When Veterans Come Home: A Response to Scott Beauchamp

PHIL KLAY

On Scott Beauchamp's essay "Did You Kill Anyone?" in *Plough*'s Spring 2021 issue, adapted from his book of the same name (Zero Books):

O NE OF THE MOST INADVERTENTLY sad TV shows of all time has to be *The West Point Story*, a cheesy 1950s Gene Roddenberry (yes, *that* Gene Roddenberry) production explicitly created, as a handsome West Point cadet tells the viewer at the beginning of each episode, to inspire pride in American citizens for one of their great institutions – the US Military Academy. The program shows a set of uniformly white, well-coiffed cadets learning lessons about honor and brotherhood and integrity (or sometimes, as in an episode actually titled "The Right to Choose," women learn that it's honorable to be an Army wife).

Viewed in 2021, it's impossible not to think of the ten- and eleven-year-old boys who watched it and imbibed its promises of a shared world of tradition and honor and ethical conduct, only to grow up, attend West Point, head to Vietnam, and return to a civilian life with that sense of a shared world utterly shattered.

In Scott Beauchamp's essay "Did You Kill Anyone?" (the opening of his brilliant book by the same name), we see Scott as a returned veteran from a different war, also in search of some kind of shared world – a community with a sufficiently common understanding of certain customs, moral attitudes, traditions, rituals, and history that the soldier's relation to US society need not be renegotiated with each conversation, a community that might enable an authentic encounter. Instead, he gets the rote, repetitive questions veterans receive: "Did you kill anyone?" "Why did you join?" (I used to joke that I should get my answer to that last one printed on a business card, so I could just hand it out at parties and spare myself explaining it yet again.)

Most poignant to me is his discussion of what he calls "Midwestern Logistical Small Talk," in which the conversation revolves around basic logistical data – have you eaten? did you have a good flight? – making "language into a comfortable and familiar meeting place where facts beget facts and everyone has equal access," enabling "a pragmatic type of communion."

But the shared world created by Midwestern Logistical Small Talk can't be deployed amid the hipster-speak he encounters in Brooklyn, which back in the late 2000s was still the pivotal node of a slowly vanishing world of apolitical aestheticism (now replaced by an attachment to fashionable politics as earnest as the earlier attachment to MGMT and The

David Modell, *Martin*, from the "Battle Scarred" portrait series

---

Phil Klay, a veteran of the US Marine Corps, published his debut novel Missionaries *in 2020 (Penguin). His short story collection* Redeployment *won the 2014 National Book Award for Fiction.*

Strokes). For this crowd, no amount of logistical data will explain Scott's time in the Army.

Partly this is a matter of geography, as Scott notes, but it's also a matter of class. Around the time I left the service, the richest twenty percent of zip codes in America sent the smallest number of kids to the military (the poorest twenty percent were also underrepresented – recruiters claim the ravages of poverty make more poor kids ineligible). The conversations Scott so ably mocks are, undoubtedly, mostly with the children of privilege for whom America is supposed to be a land of opportunities (and in those years specifically opportunities for self-expression), not sacrifices.

Against this rather depressing backdrop, conservatives sometimes like to contrast an earlier time in which there really was a shared world, the American fantasy so cheerily portrayed in *The West Point Story*. But the cheerful world of the show, in which the divide between citizen and soldier was bridged through the careful articulation of a sanitized vision of service (though it did capture something true about the aspirations of the Army), was a deliberate historical creation, one that eventually, violently, divided veterans against each other.

Reading Scott's essay and reflecting on his interlocutors' confusion about the meaning of his service, I thought not only of *The West Point Story*, but about a real event in the aftermath of World War I. As Jonathan Ebel details in his excellent *GI Messiahs*, a 1919 Armistice Day parade led by members of the American Legion ended up raiding the Centralia, Washington, offices of the Industrial Workers of the World and getting into an exchange of gunfire.

The Legion had been formed that year to promote a patriotic and collective vision of America that would keep alive "the spirit of the Great War" and protect "the American way." Such a collective vision, though, is at odds with the other promise of America, that of pluralism – hence the attitude expressed by Chicago-based Legionnaires that "we still have a fight to be carried to a victorious conclusion here at home."

That fight continued in a very real sense at the Industrial Workers of the World offices, where one of the Wobblies, as they were called, was a man named Wesley Everest. Everest was a veteran of the Great War as well, but his reaction to the "spirit of the Great War" was to join a revolutionary industrial-union movement, rather than one dedicated to cultivating civil and religious orthodoxy and deep reverence for a mythic past. And when violence broke out at the Legion parade, he took up arms and, supposedly, shot two of the rioting Legionnaires – men who were his ideological opposites but also fellow former soldiers. He was arrested, along with several others, but a lynch mob formed and broke into the jail. They left most of the Wobblies alone, but Everest was a veteran, an apostate. They dragged him out under the cover of darkness, beat him, hanged him, and shot him repeatedly before dragging him back to his cell and tossing his dead body in with the living prisoners.

An editorial published in the *New York Times* a few days later found a silver lining in the event, insofar as it could be used to further glorify the American Legion and their mission:

> The word "martyr" is often misused and applied to men who are merely victims. A martyr is one whose death is caused by his support of principles and convictions. The dead soldiers who were marked for slaughter by the I.W.W. because they believed in the American flag were martyrs to that belief, and the blood of the martyrs is the seed of the church.

When ideological divisions are too stark, and Midwestern Logistical Small Talk isn't enough to fill the gap, that's how a shared world is made. ⤙

For an urban beekeeper
in Manhattan, the hive is
a city within a city.

TIM MAENDEL

City of Bees

I AM LOOKING FOR EGGS. Not the hard-shelled ones you cook for breakfast; these are very small, half the size of a grain of rice. It would be great to meet the queen, but I am satisfied that she's close – I see evidence of her in her subjects' focus, the calm her presence brings. I look for baby food – not Gerber purées, but pollen packed into storage cells. And of course, I'm checking for honey. I have escaped the world and am inside a beehive.

I'm an intruder here and must watch my step. A wrong or hasty move can set off an angry defense that I have learned to regret. But while things are going well I'm loving my temporary immersion in nature.

An astronaut begins his journey when he climbs into a spacesuit and has a partner screw on his helmet. Mine starts when I flip down my veil and secure the seams. My launch comes when I pry off the lid of the hive, the loud crack

The author
tending beehives
on a New York
City rooftop

Tim Maendel lives in the Bruderhof's Harlem House, New York City, with his wife, his dog, and forty-five thousand bees.

a barn, it's always my entry point to a harsh and beautiful world.

It is not my world. Despite being called a "keeper" I have no control here. I am an observer, capable perhaps of small assistance for the needs I see, or of compensating for the limitations of the manmade boundaries that I put them into. I don't understand half of what is going on; I am often reminded of that. I can discover a problem only to find the bees are already halfway to fixing it themselves. I have seen a queenless hive, doomed to fail, and rushed out to buy a queen – only to find on my return that the bees were well on the

of the broken propolis (bee glue) seal standing in for the boom of booster engines.

It doesn't matter where the beehive is located; the mini-world I enter is always the same. Sometimes hives are in lonely, grassy fields, with wildlife hidden in nearby woods. At other times, I ride elevators up skyscrapers, climb stairs, and walk through rooms full of whining machines, out onto flat roofs high above city boardrooms full of suit-wearing executives. New York City is all around me, the Hudson River far below. Elsewhere, lower down, I step out of apartments into rooftop gardens where the streaks of racing bees point me toward the hive. But whether I pry off the lid next to the corporate offices of a beauty-product company, a suburban mansion, or

way to making their own, feeding a larva with the special food that transforms her. In fact, there have been times when the queen I bring is rejected and killed by the one they make. I've baited empty hives with ready-to-use comb and tempting honey right by a fleeing swarm, desperate to catch them, and seen it ignored. If ever I feel important to them, I soon remember that they don't know me.

The concentration bee-watching requires seems to free other parts of my mind for creativity. Solutions to issues I didn't know I was even thinking about, inspirations and mini-resolutions have suddenly presented themselves to my mind while I'm in a hive. The balance of wonder and danger energizes my thoughts. Sometimes the humming cloud

I work in seems friendly, like I am being welcomed as a temporary co-worker on a sixty-thousand-member team. At other times the angry buzz of bees bouncing off my veil and gloves reminds me of the deadly power that I've intruded on. The ones that manage to sting me punctuate the fact of their power to kill me. I am glad for every square inch of protection.

A hive is a super-organism that makes single decisions powered by tens of thousands of individuals. Each bee has a specific role – foragers work so hard collecting nectar and pollen from a three- to five-mile radius during the spring, summer, and fall that their one-month lifetimes are less than a fifth of those of winter bees. There are nurse bees, cleaners, food processors, and guards who also take care of the hive's temperature. When it overheats they go on fan duty at the entrance, planting their feet and revving their wings up to full flight thrust to push in fresh air. A few drones get to mate with the queen – and die immediately afterwards; the rest seem to wander around the hive all summer before being kicked out. Guard bees are ready to give their lives, protecting others with a venomous sting.

But this high level of specialization in roles is also what constitutes the tight body made up of all the bees. A body that makes one decision, has one health, and makes a product that is replenished, apparently without complaint, after we take what we decide is our share. Each hive seems to have a character, its own micro-cosmic *Volksgeist*. It can turn on me suddenly, and then change in response to the smoke I puff, and turn away. One hive can work hard at making honey when a hive right next to it, started on the same day, makes hardly any. A hive can be so friendly one day that I wonder if they even notice I am there, working alongside them. On another, especially after a wrong move that startles or threatens them, some signal seems to go out and bees are all around, filling my head with their buzzing and following me away from the hive, waiting to sting after I take off the veil.

Recently, I delivered honey from a client's hives to his home. When I was leaving he walked out with me. "Thanks again for the honey," he said, tapping me on the shoulder for emphasis. "You don't know what this means to us." But I understood that this household in the middle of a large city had just connected with nature. "You are a farmer!" I told him.

Philip Britts – writer, poet, pastor, and visionary observer of his natural surround-ings – speaks of the soul-deepening value of being close to nature and even warns that the loss of connection to the simplicity and faith of rural life leads to the loss of "inner stability." Britts wrote ten points that define a "good farmer." Here's the last one, which I find both humbling and deepening. A good farmer, he says, "realizes that he knows next to nothing of all that there is to know, that he is dealing with eternal laws which he did not make and cannot alter, and that the most brilliant achievements of human knowledge are simply the closest obedience to these laws."

Like the urban-farmer client I'm assisting, I am thankful for any such connection to eternal laws. I also enjoy the taste of the rewards, and hope to be a part of bringing them to many more. ⟶

> The beehive is not my world. Despite being called a "keeper" I have no control here.

Sister Dorothy Stang

SUSANNAH BLACK AND JASON LANDSEL

"IF ANYTHING HAPPENS," her friend Ivan remembers her saying that weekend, "I hope it happens to me; the others have families."

Sister Dorothy Stang arrived in the Brazilian rainforest around the same time as the agribusiness tycoons. In 1966, when she and four other Sisters of Notre Dame de Namur were sent to help forest-farming peasants establish economic self-reliance, the government had just begun its romance with the World Bank to develop the Amazon for logging, mining, and cattle ranching. The *Wall Street Journal* called it the Brazilian Miracle. The peasants called it *capitalismo selvagem,* a savage wild beast.

Dorothy had been raised Catholic in Dayton, Ohio. At seventeen, she and her best friend entered the same religious order. She had a deep desire to serve the poor and befriend people very different from herself.

When she arrived in Brazil, it soon became clear that serving the poor was not going to be an apolitical proposition. It was a pivotal era, with Latin American bishops boldly applying the teachings of the gospel to local injustices. Dorothy helped establish dozens of the "base ecclesial communities" that were springing up throughout Latin America: lay-led cooperatives for both spiritual development and economic empowerment. She learned Portuguese and local indigenous languages, set up schools, and repeatedly, persistently, filed legal claims on behalf of small farmers whose land was being stolen.

She also was concerned to preserve the forests themselves, at a time when the environmental movement was in its infancy. Her favorite T-shirt read "The death of the forest is the end of our lives." Part of her work was advocating for reforesting stripped areas with appropriate crop-trees such as coffee and açaí, and encouraging small farmers to use intensive techniques rather than the slash-and-burn practices of the big landowners.

By the late 1990s, deforestation was coming under greater scrutiny, and a group of ranchers and miners put Dorothy's name at the top of a death list of those who were making too much trouble for them.

On February 12, 2005, Dorothy was up early and on her way to another community meeting to strategize a response to illegal logging. She had heard rumors that a few ranchers and loggers had met in January to determine how to have her killed and who would pay the assassins. Things had been heating up. Gunmen had just driven Luis, a small farmer whose land a rancher wanted, from his home and burned it down. When she got to the meeting, Dorothy would organize a place for the displaced family to stay.

As she got to the top of a low hill, under the forest canopy, two men blocked her path. She knew them; they worked for the man who had driven Luis out. "Do you have a weapon?" they asked.

"Yes," she answered, and pulled out her tattered Bible. She opened it and read: "Blessed are the poor in spirit. Blessed are those who hunger and thirst for justice. Blessed are the peacemakers." She looked at the men standing in her way. "God bless you, my sons."

They shot her six times.

The obituaries said things like, "She gave her life for her causes," evoking a do-gooder with a collection of worthy bees in her bonnet. But no. Already as a teen, she gave her life as a handmaid of Christ, king of a kingdom where all people will live in harmony with each other and the earth. ⤳

Susannah Black is a Plough *editor. Jason Landsel is the artist for* Plough's *"Forerunners" series.*